Speed Up Your Python with Rust

Optimize Python performance by creating Python pip modules in Rust with PyO3

Maxwell Flitton

BIRMINGHAM—MUMBAI

Speed Up Your Python with Rust

Group Product Manager: Richa Tripathi

Publishing Product Manager: Richa Tripathi

Senior Editor: Nisha Cleetus

Content Development Editor: Vaishali Ramkumar

Technical Editor: Pradeep Sahu

Copy Editor: Safis Editing

Project Coordinator: Deeksha Thakkar

Proofreader: Safis Editing

Indexer: Manju Arasan

Production Designer: Shankar Kalbhor

First published: December 2021

Production reference: 1151221

Published by Packt Publishing Ltd.
Livery Place
35 Livery Street
Birmingham
B3 2PB, UK.

ISBN 978-1-80181-144-6

www.packt.com

To my wife, Melanie Zhang, who stuck with me and supported me through a busy work schedule and deadlines. Not only are you smart and caring, you have been an amazing team player.

– Maxwell Flitton

Contributors

About the author

Maxwell Flitton is a software engineer who works for the open source financial loss modeling foundation OasisLMF. In 2011, Maxwell achieved his Bachelor of Science degree in nursing from the University of Lincoln, UK. While working 12-hour shifts in the A&E departments of hospitals, Maxwell obtained another degree in physics from the Open University in the UK and then moved on to another milestone, with a postgraduate diploma in physics and engineering in medicine from UCL in London. He's worked on numerous projects such as medical simulation software for the German government and supervising computational medicine students at Imperial College London. He also has experience in financial tech and Monolith AI.

Many thanks to the Rust community for developing an amazing language with a friendly community that's willing to push boundaries. I'm also grateful to the team at Monolith AI, where Saravanan Sathyanandha and Richard Ahlfeld empowered me to grow as an engineer. This has been carried further by the OasisLMF team where Ben Hayes, Stephane Struzik, Sam Gamble, and Hassan Chagani have been supportive and enabled me to grow.

About the reviewers

Mário Idival is a Brazilian and a lover of technologies aimed at software development, mainly focused on programming languages. He acts as a technical manager and software engineer in his spare time. He started his journey as a software developer in 2011 learning only with instructional material from the internet, starting with learning C, and soon after switched to Python, which allowed him to achieve his first job after 6 months of studies.

Today, with 10 years of experience, he has delivered software in several areas including loans, tourism and travel, artificial intelligence, electronic data interchange, process automation, and cryptocurrency. He is currently focused on learning and spreading knowledge of the Rust language. He also supports the Rust community in the Rust By Example project.

Boyd Johnson has been working in software since 2015. As part of a team at Bitwise IO, along with partners at Intel, he worked to develop Hyperledger Sawtooth, an open source blockchain, in Python and Rust. Boyd worked, in particular, on the FFI layer between Python and Rust, as well as transaction processing components. You can read more of Boyd's writing at `boydjohnson.dev`.

Table of Contents

Preface

Section 1: Getting to Understand Rust

1

An Introduction to Rust from a Python Perspective

Technical requirements	4	Move	23
Understanding the differences		Immutable borrow	24
between Python and Rust	5	Mutable borrow	26
Why fuse Python with Rust?	5	**Keeping track of scopes**	
Passing strings in Rust	8	**and lifetimes**	**26**
Sizing up floats and integers in Rust	10	**Building structs instead**	
Managing data in Rust's vectors and		**of objects**	**30**
arrays	12	**Metaprogramming with macros**	
Replacing dictionaries with hashmaps	14	**instead of decorators**	**34**
Error handling in Rust	18	Summary	37
		Questions	38
Understanding variable		Answers	38
ownership	**21**	Further reading	39
Copy	22		

2

Structuring Code in Rust

Technical requirements	42	Structuring code over multiple	
Managing our code with crates		files and modules	50
and Cargo instead of pip	42	Building module interfaces	55

Benefits of documentation
when coding   64
Interacting with the
environment   65

Summary   68
Questions   68
Answers   69
Further reading   70

3
Understanding Concurrency

Technical requirements   72
Introducing concurrency   72
Threads   72
Processes   74

Basic asynchronous
programming with threads   74
Running multiple processes   82
Customizing threads and
processes safely   95

Amdahl's law   95
Deadlocks   96
Race conditions   98

Summary   99
Questions   99
Answers   100
Further reading   101

Section 2: Fusing Rust with Python

4
Building pip Modules in Python

Technical requirements   105
Configuring setup tools for
a Python pip module   106
Creating a GitHub repository   106
Defining the basic parameters   109
Defining a README file   111
Defining a basic module   111

Packaging Python code in
a pip module   113
Building our Fibonacci calculation code 114
Creating a command-line interface   116
Building unit tests   118

Configuring continuous
integration   125
Manually deploying onto PyPI   126
Managing dependencies   127
Setting up type checking for Python 129
Setting up and running tests and
type-checking with GitHub Actions   131
Create automatic versioning for our
pip package   135
Deploying onto PyPI using GitHub
Actions   138

Summary   141

Questions   142
Answers   142

Further reading   143

5

Creating a Rust Interface for Our pip Module

Technical requirements   146
Packaging Rust with pip   146
Define gitignore and Cargo for our
package   147
Configuring the Python setup process
for our package   149
Installing our Rust library for our
package   151

Building a Rust interface with
the pyO3 crate   153
Building our Fibonacci Rust code   153

Creating command-line tools for
our package   157
Creating adapters for our package   159

Building tests for our Rust
package   167
Comparing speed with Python,
Rust, and Numba   170
Summary   173
Questions   173
Answers   174
Further reading   174

6

Working with Python Objects in Rust

Technical requirements   176
Passing complex Python
objects into Rust   176
Updating our setup.py file to support
.yml loading   177
Defining our .yml loading command   178
Processing data from our Python
dictionary   179
Extracting data from our config file   183
Returning our Rust dictionary to our
Python system   185

Inspecting and working with
custom Python objects   187
Creating an object for our Rust
interface   188

Acquiring the Python GIL in Rust   189
Adding data to our newly created
PyDict struct   191
Setting the attributes of our custom
object   193

Constructing our own custom
Python objects in Rust   195
Defining a Python class with the
required attributes   195
Defining class static methods to
process input numbers   196
Defining a class constructor   197
Wrapping up and testing our module   198

Summary   202

| Questions | 202 | Further reading | 203 |
| Answers | 203 | | |

7

Using Python Modules with Rust

Technical requirements	206	Building get_weight_matrix and inverse_weight_matrix functions	223
Exploring NumPy	206		
Adding vectors in NumPy	206	Building get_parameters, get_times, and get_input_vector functions	224
Adding vectors in pure Python	208		
Adding vectors using NumPy in Rust	209	Building calculate_parameters and calculate_times functions	226
Building a model in NumPy	213	Adding calculate functions to the Python bindings and adding a NumPy dependency to our setup.py file	227
Defining our model	213		
Building a Python object that executes our model	216	Building our Python interface	228
Using NumPy and other Python modules in Rust	219	Summary	229
		Questions	230
Recreating our NumPy model in Rust	222	Answers	231
		Further reading	231

8

Structuring an End-to-End Python Package in Rust

Technical requirements	234	Utilizing and testing our package	252
Breaking down a catastrophe modeling problem for our package	234	Building a Python construct model using pandas	253
Building an end-to-end solution as a package	239	Building a random event ID generator function	254
Building a footprint merging process	240	Timing our Python and Rust implementations with a series of different data sizes	255
Building the vulnerability merge process	243		
Building a Python interface in Rust	247	Summary	257
Building our interface in Python	249	Further reading	258
Building package installation instructions	250		

Section 3: Infusing Rust into a Web Application

9
Structuring a Python Flask App for Rust

Technical requirements	262	Setting up the application database migration system	280
Building a basic Flask application	262	Building database models	283
Building an entry point for our application	263	Applying the database access layer to the fib calculation view	285
Building our Fibonacci number calculator module	264	**Building a message bus**	**287**
Building a Docker image for our application	266	Building a Celery broker for Flask	288
Building our NGINX service	268	Building a Fibonacci calculation task for Celery	290
Connecting and running our Nginx service	270	Updating our calculation view	291
		Defining our Celery service in Docker	292
Defining our data access layer	**273**	**Summary**	**295**
Defining a PostgreSQL database in docker-compose	275	**Questions**	**296**
Building a config loading system	275	**Answers**	**296**
Building our data access layer	277	**Further reading**	**297**

10
Injecting Rust into a Python Flask App

Technical requirements	300	Deploying Flask and Celery with Rust	307
Fusing Rust into Flask and Celery	300	Deploying with a private GitHub repository	308
Defining our dependency on the Rust Fibonacci number calculation package	301	Building a Bash script that orchestrates the whole process	310
Building our calculation model with Rust	301	Reconfiguring the Rust Fib package installment in our Dockerfile	311
Creating a calculation view using Rust	304		
Inserting Rust into our Celery task	305	**Fusing Rust with data access**	**312**

Setting up our database cloning
package   313
Setting up the diesel environment   315
Autogenerating and configuring our
database models and schema   317
Defining our database connection
in Rust   319

Creating a Rust function that
gets all the Fibonacci records
and returns them   320

Deploying Rust nightly in Flask   322
Summary   324
Questions   324
Answers   325
Further reading   325

11
Best Practices for Integrating Rust

Technical requirements   328
**Keeping our Rust
implementation simple by
piping data to and from Rust   328**
Building a Python script that
formulates the numbers for
calculation   329
Building a Rust file that accepts the
numbers, calculates the Fibonacci
numbers, and returns the calculated
numbers   330
Building another Python script that
accepts the calculated numbers and
prints them out   331

**Giving the interface a native
feel with objects   334**
Defining traits   339
Defining struct behavior with traits   341
Passing traits through functions   344
Storing structs with common traits   345
Running our traits in the main file   346

**Keeping data-parallelism
simple with Rayon   349**
Further reading   351

Index
Other Books You May Enjoy

Preface

The Rust programming language is an exciting new language. It gives us memory safety without garbage collection, resulting in fast times and low memory footprints. However, rewriting everything in Rust can be expensive and risky as there might not be package support in Rust for the problem being solved. This is where Python bindings and pip come in. This book will enable you to code modules in Rust that can be installed using pip. As a result, you will be able to inject Rust as and when you need it without taking on the risk and workload of rewriting your entire system. This enables you, as a developer, to experiment with and use Rust in your Python projects.

Who this book is for

Python developers who want to speed up their code with Rust, or experiment with Rust without having to take on much risk or workload, will benefit from this book. No background in Rust is needed. This book has an introduction to Rust for Python developers, and uses Python examples to get you up to speed with Rust quickly.

What this book covers

Chapter 1, *An Introduction to Rust from a Python Perspective*, covers the basics of Rust to enable Rust development. Relevant Python examples are given to help you grasp the Rust concepts examined.

Chapter 2, *Structuring Code in Rust*, explains how to structure Rust programs over multiple pages and use package management tools to organize and install dependencies.

Chapter 3, *Understanding Concurrency*, covers how to multithread and multiprocess in Rust, seeing as Rust has "fearless concurrency." We also cover concurrency in Python to see the differences.

Chapter 4, *Building pip Modules in Python*, sees us build Python packages that can be installed using pip. It also covers how packages can be hosted privately on GitHub.

Chapter 5, *Creating a Rust Interface for Our pip Module*, has us inject Rust into our pip module and use the Rust setup tools to compile and use the Rust code in our Python code.

Chapter 6, Working with Python Objects in Rust, considers how compatibility does not just go in one direction. In this chapter, we take in Python objects and interact with them. We also create Python objects in Rust.

Chapter 7, Using Python Modules in Rust, builds on the previous chapter and sees us use Python modules such as NumPy in our Rust code.

Chapter 8, Structuring an End-to-End Python Module in Rust, sees us wrapping up everything that has been covered into a fully functioning Python package written in Rust. This package has Python interfaces and command-line functionality that accepts YAML files for configuration.

Chapter 9, Structuring a Python Flask App for Rust, has us build a Python Flask app with a PostgreSQL database, NGINX load balancer, and Celery worker in order to get more practical with our Rust skills. All of this is wrapped in Docker to prepare us for injecting Rusk into all of these aspects of the web application.

Chapter 10, Injecting Rust into a Python Flask App, covers how to take the web application that we built in the previous chapter and inject our Rust modules into the Docker containers for the Celery worker and Flask application. We also imprint the migrations that have already been applied to automatically generate a schema of the database so our Rust code can directly connect with the database.

Chapter 11, Best Practices for Integrating Rust, concludes the book with some tips on how to avoid common mistakes as you continue to write Rust code for Python.

To get the most out of this book

It is advisable that you understand Python and are comfortable with object-Oriented programming. Some advanced topics such as meta-classing will be touched on but are not essential. Rust programming, Python web apps, and Python modules installed using pip are all covered in the book.

Software/hardware covered in the book	Operating system requirements
Python 3	Windows, macOS, or Linux
Rust	Windows, macOS, or Linux
Docker	Windows, macOS, or Linux
Py03	Windows, macOS, or Linux
Redis	Windows, macOS, or Linux
PostgreSQL	Windows, macOS, or Linux

If you are using the digital version of this book, we advise you to type the code yourself or access the code from the book's GitHub repository (a link is available in the next section). Doing so will help you avoid any potential errors related to the copying and pasting of code.

Download the example code files

You can download the example code files for this book from GitHub at `https://github.com/PacktPublishing/Speed-up-your-Python-with-Rust`. If there's an update to the code, it will be updated in the GitHub repository.

We also have other code bundles from our rich catalog of books and videos available at `https://github.com/PacktPublishing/`. Check them out!

Download the color images

We also provide a PDF file that has color images of the screenshots and diagrams used in this book. You can download it here: `https://static.packt-cdn.com/downloads/9781801811446__ColorImages.pdf`.

Conventions used

There are a number of text conventions used throughout this book.

`Code in text`: Indicates code words in text, database table names, folder names, filenames, file extensions, pathnames, dummy URLs, user input, and Twitter handles. Here is an example: "Mount the downloaded `WebStorm-10*.dmg` disk image file as another disk in your system."

A block of code is set as follows:

```
use std::error::Error;
use std::fs::File;
use csv;

use super::structs::FootPrint;
```

When we wish to draw your attention to a particular part of a code block, the relevant lines or items are set in bold:

```
let code = "5 + 6";
let result = py.eval(code, None, Some(&locals)).unwrap();
let number = result.extract::<i32>().unwrap();
```

Any command-line input or output is written as follows:

```
pip install git+https://github.com/maxwellflitton/flitton-
fib-rs@main
```

Bold: Indicates a new term, an important word, or words that you see on screen. For instance, words in menus or dialog boxes appear in **bold**. Here is an example: "This can be done by clicking on the **Settings** tab and then the **Secrets** tab on the left sidebar, as seen here."

> **Tips or Important Notes**
> Appear like this.

Get in touch

Feedback from our readers is always welcome.

General feedback: If you have questions about any aspect of this book, email us at customercare@packtpub.com and mention the book title in the subject of your message.

Errata: Although we have taken every care to ensure the accuracy of our content, mistakes do happen. If you have found a mistake in this book, we would be grateful if you would report this to us. Please visit www.packtpub.com/support/errata and fill in the form.

Piracy: If you come across any illegal copies of our works in any form on the internet, we would be grateful if you would provide us with the location address or website name. Please contact us at copyright@packt.com with a link to the material.

If you are interested in becoming an author: If there is a topic that you have expertise in and you are interested in either writing or contributing to a book, please visit authors.packtpub.com.

Share Your Thoughts

Once you've read *Speed Up your Python with Rust*, we'd love to hear your thoughts! Scan the QR code below to go straight to the Amazon review page for this book and share your feedback.

https://packt.link/r/1-801-81144-X

Your review is important to us and the tech community and will help us make sure we're delivering excellent quality content.

Section 1: Getting to Understand Rust

In this section, we will get to grips with Rust. Instead of introducing the basics of Rust, such as loops and functions, we will cover the syntax specific to Rust. After this, we will explore the quirks that the Rust language introduces, primarily centered around memory management. We will then cover how to manage dependencies and structure our code over multiple files. After this, we will experiment with multithreading and multiprocessing in Rust and Python.

This section comprises the following chapters:

- *Chapter 1, An Introduction to Rust from a Python Perspective*
- *Chapter 2, Structuring Code in Rust*
- *Chapter 3, Understanding Concurrency*

1

An Introduction to Rust from a Python Perspective

Due to its speed and safety, it is no surprise that **Rust** is the new language gaining in popularity. However, with success comes criticism. Despite Rust's popularity as an impressive language, it has also gained the label of being hard to learn, an idea which isn't quite grounded in reality.

In this chapter, we will cover all of Rust's quirks that will be new to a Python developer. If Python is your main language, concepts such as basic memory management and typing can initially slow down your ability to quickly write productive Rust code due to the compiler failing to compile the code. However, this can quickly be overcome by learning the rules around Rust features, such as variable ownership, lifetimes, and so on, as Rust is a memory-safe language. Consequently, we must keep track of our variables as they usually get deleted instantly when they go out of scope. If this does not make sense yet, don't worry; we will cover this concept in the *Keeping track of scopes and lifetimes* section.

In this chapter, we will also be covering the basics of syntax, while you will be setting up a Rust environment on your own computer in the next chapter. Do not worry though, you can code all the examples in this chapter on the free online Rust playground.

In particular, we will cover the following topics in this chapter:

- Understanding the differences between Python and Rust
- Understanding variable ownership
- Keeping track of scopes and lifetimes
- Building structs as opposed to objects
- Metaprogramming with macros instead of decorators

Technical requirements

As this is just an introduction, all the Python examples in the chapter can be implemented with a free online Python interpreter such as `https://replit.com/languages/python3`.

The same goes for all the Rust examples. These can be implemented using the free online Rust playground found at `https://play.rust-lang.org/`.

The code covered in the chapter can be found at `https://github.com/PacktPublishing/Speed-up-your-Python-with-Rust/tree/main/chapter_one`.

Understanding the differences between Python and Rust

Rust can sometimes be described as a *systems language*. As a result, it can sometimes be labeled by software engineers in a way that is similar to C++: fast, hard to learn, dangerous, and time-consuming to code in. As a result, most of you mainly working in dynamic languages such as Python could be put off. However, Rust is memory-safe, efficient, and productive. Once we have gotten over some of the quirks that Rust introduces, nothing is holding you back from exploiting Rust's advantages by using it to write fast, safe, and efficient code. Seeing as there are so many advantages to Rust, we will explore them in the next section.

Why fuse Python with Rust?

When it comes to picking a language, there is usually a trade-off between resources, speed, and development time. Dynamic languages such as Python became popular as computing power increased. We were able to use the extra resources we had to manage our memory with garbage collectors. As a result, developing software became easier, quicker, and safer. As we will cover later in the *Keeping track of scopes and lifetimes* section, poor memory management can lead to some security flaws. The exponential increase in computing power over the years is known as **Moore's Law**. However, this is not continuing to hold and in 2019, Nvidia's CEO Jensen Huang suggested that as chip components get closer to the size of individual atoms, it has gotten harder to keep up with the pace of Moore's Law, thus declaring it dead (`https://www.cnet.com/news/moores-law-is-dead-nvidias-ceo-jensen-huang-says-at-ces-2019/`).

However, with the rise of big data, our need to pick up faster languages to satisfy our needs is increasing. This is where languages such as Golang and Rust enter. These languages are memory-safe, yet they compile and have significant speed increases. What makes Rust even more unique is that it has managed to achieve memory safety without **garbage collection**. To appreciate this, let's briefly describe garbage collection: this is where the program temporarily stops, checks all the variables to see which ones are no longer being used, and deletes those that are not. Considering that Rust does not have to do this, it is a significant advantage as Rust does not have to keep stopping to clean up the variables. This was demonstrated in Discord's 2020 blog post *Why Discord is switching from Go to Rust*: `https://blog.discord.com/why-discord-is-switching-from-go-to-rust-a190bbca2b1f#:~:text=The%20service%20we%20switched%20from,is%20in%20the%20hot%20path`. In this post, we can see that Golang just could not keep up with Rust, as demonstrated in the graph they presented:

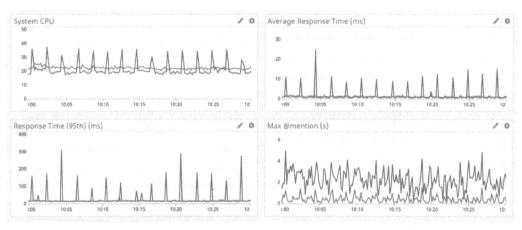

Figure 1.1 – Golang is spiky and Rust is the flat line below Golang (image source: https://blog.discord. com/why-discord-is-switching-from-go-to-rust-a190bbca2b1f#:~:text=The%20service%20we%20 switched%20from,is%20in%20the%20hot%20path)

The comments on the post were full of people complaining that Discord used an out-of-date version of Golang. Discord responded to this by stating that they tried a range of Golang versions, and they all had similar results. With this, it makes sense to get the best of both worlds without much compromise. We can use Python for prototyping and complex logic. The extensive range of third-party libraries that Python has combined with the flexible object-oriented programming it supports make it an ideal language for solving real-world problems. However, it's slow and is not efficient with the use of resources. This is where we reach for Rust.

Rust is a bit more restrictive in the way we can lay out and structure the code; however, it's fast, safe, and efficient when implementing multithreading. Combining these two languages enables a Python developer to have a powerful tool in their belt that their Python code can use when needed. The time investment needed to learn and fuse Rust is low. All we must do is package Rust and install it in our Python system using pip and understand a few quirks that Rust has that are different from Python. We can start this journey by looking at how Rust handles strings in the next section. However, before we explore strings, we have to first understand how Rust is run compared to Python.

If you have built a web app in Python using Flask, you will have seen multiple tutorials sporting the following code:

```python
from flask import Flask
app = Flask(__name__)


@app.route("/")
def home():
    return "Hello, World!"


if __name__ == "__main__":
    app.run(debug=True)
```

What we must note here is the last two lines of the code. Everything above that defines a basic Flask web app and a route. However, the running of the app in the last two lines only executes if the Python interpreter is directly running the file. This means that other Python files can import the Flask app from this file without running it. This is referred to by many as an *entry point*.

You import everything you need in this file, and for the application to run, we get our interpreter to run this script. We can nest any code under the if __name__ == "__main__": line of code. It will not run unless the file is directly hit by the Python interpreter. Rust has a similar concept. However, this is more essential, as opposed to Python that just has it as a nice-to-have feature. In the Rust playground (see the *Technical requirements* section), we can type in the following code if it is not there already:

```rust
fn main() {
    println!("hello world");
}
```

This is the entry point. The Rust program gets compiled, and then runs the `main` function. If whatever you've coded is not accessed by the `main` function, it will never run. Here, we are already getting a sense of the safety enforced by Rust. We will see more of this throughout the book.

Now that we have our program running, we can move on to understanding the difference between Rust and Python when it comes to strings.

Passing strings in Rust

In Python, **strings** are flexible. We can pretty much do what we want with them. While technically, Python strings cannot be changed under the hood, in Python syntax, we can chop and change them, pass them anywhere, and convert them into integers or floats (if permitted) without having to think too much about it. We can do all of this with Rust too. However, we must plan beforehand what we are going to do. To demonstrate this, we can dive right in by making our own `print` function and calling it, as seen in the following code:

```
fn print(input: str) {
    println!("{}", input);
}
fn main() {
    print("hello world");
}
```

In Python, a similar program would work. However, when we run it in the Rust playground, we get the following error:

```
error[E0277]: the size for values of type 'str' cannot be known
at compilation time
```

This is because we cannot specify what the maximum size is. We don't get this in Python; therefore, we must take a step back and understand how variables are assigned in memory. When the code compiles, it allocates memory for different variables in the stack. When the code runs, it stores data in the heap. Strings can be various sizes so we cannot be sure at compile time how much memory we can allocate to the `input` parameter of our function when compiling. What we are passing in is a **string slice**. We can remedy this by passing in a string and converting our string literal to a string before passing it into our function as seen here:

```
fn print(input: String) {
    println!("{}", input);
```

```
}
fn main() {
    let string_literal = "hello world";
    print(string_literal.to_string());
}
```

Here, we can see that we have used the `to_string()` function to convert our string literal into a string. To understand why `String` is accepted, we need to understand what a string is.

A string is a type of wrapper implemented as a vector of bytes. This vector holds a reference to a string slice in the heap memory. It then holds the amount of data available to the pointer, and the length of the string literal. For instance, if we have a string of the string literal *one*, it can be denoted by the following diagram:

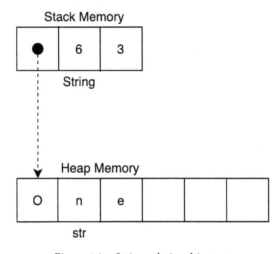

Figure 1.2 – String relationship to str

Considering this, we can understand why we can guarantee the size of `String` when we pass it into our function. It will always be a pointer to the string literal with some meta-information about the string literal. If we can just make a reference to the string literal, we can pass this into our function as it is just a reference and we can therefore guarantee that the size of the reference will stay the same. This can be done by borrowing using the & operator as shown in the following code:

```
fn print(input_string: &str) {
    println!("{}", input_string);
}
fn main() {
```

```
    let test_string = &"Hello, World!";
    print(test_string);
}
```

We will cover the concept of borrowing later in the chapter but, for now, we understand that, unlike Python, we must guarantee the size of the variable being passed into a function. We can use borrowing and wrappers such as strings to handle this. It may not come as a surprise, but this does not just stop at strings. Considering this, we can move on to the next section to understand the differences between Python and Rust when it comes to floats and integers.

Sizing up floats and integers in Rust

Like strings, Python manages floats and integers with ease and simplicity. We can pretty much do whatever we want with them. For instance, the following Python code will result in 6.5:

```
result = 1 + 2.2
result = result + 3.3
```

However, there is a problem when we try to just execute the first line in Rust with the following line of Rust code:

```
let result = 1 + 2.2;
```

It results in an error telling us that a float cannot be added to an integer. This error highlights one of the pain points that Python developers go through when learning Rust, as Rust enforces typing aggressively by refusing to compile if typing is not present and consistent. However, while this is an initial pain, aggressive typing does help in the long run as it maintains safety.

Type annotation in Python is gaining popularity. This is where the type of the variable is declared for parameters of functions or variables declared, enabling some editors to highlight when the types are inconsistent. The same happens in JavaScript with TypeScript. We can replicate the Python code at the start of this section with the following Rust code:

```
let mut result = 1.0 + 2.2;
result = result + 3.3;
```

It has to be noted that the `result` variable must be declared as a **mutable** variable with the `mut` notation. Mutable means that the variable can be changed. This is because Rust automatically assigns all variables as immutable unless we use the `mut` notation.

Now that we have seen the effects of types and mutability, we should really explore **integers** and **floats**. Rust has two types of integers: **signed integers**, which are denoted by `i`, and **unsigned integers**, denoted by `u`. Unsigned integers only house positive numbers, whereas signed integers house positive and negative integers. This does not just stop here. In Rust, we can also denote the size of the integer that is allowed. This can be calculated by using binary. Now, understanding how to use binary notation to describe numbers in detail is not really needed. However, understanding the simple rule that the size can be calculated by raising two to the power of the number of bits can give us an understanding of how big an integer is allowed to be. We can calculate all the integer sizes that we can utilize in Rust with the following table:

Bits	Calculation	Size
8	2^8	256
16	2^16	65536
32	2^32	4294967296
64	2^64	1.8446744e+19
128	2^128	3.4028237e+38

Table 1.1 – Size of integer types

As we can see, we can get to very high numbers here. However, it is not the best idea to assign all variables and parameters as `u128` integers. This is because the compiler will set aside this amount of memory each time when compiling. This is not very efficient considering that it's unlikely that we will be using such large numbers. It must be noted that the changes in each jump are so large it is pointless graphing it. Each jump in bits completely overshadows all the others, resulting in a flat line along the x axis and a huge spike at the last graphed number of bits. However, we also must be sure that our assignment is not too small. We can demonstrate this with the Rust code as follows:

```
let number: u8 = 255;
let breaking_number: u8 = 256;
```

Our compiler will be OK with the `number` variable. However, it will throw the error shown next when assigning the `breaking_number` variable:

```
literal '256' does not fit into the type 'u8' whose range
is '0..=255'
```

This is because there are 256 integers between 0 -> 255, as we include 0. We can change our unsigned integer to a signed one with the following line of Rust code:

```
let number: i8 = 255;
```

This gives us the following error:

```
literal '255' does not fit into the type 'i8' whose range
is '-128..=127'
```

In this error, we are reminded that the bits are are allocated memory space. Therefore, an i8 integer must accommodate positive and negative integers within the same number of bits. As a result, we can only support a magnitude that is half of the integer of an unsigned integer.

When it comes to floats, our choices are more limited. Here, Rust accommodates both f32 and f64 floating points. Declaring these floating-point variables requires the same syntax as integers:

```
let float: f32 = 20.6;
```

It must be noted that we can also annotate numbers with suffixes, as shown in the following code:

```
let x = 1u8;
```

Here, x has a value of 1 with the type of u8. Now that we have covered floats and integers, we can use vectors and arrays to store them.

Managing data in Rust's vectors and arrays

With Python, we have lists. We can stuff anything we want into these lists with the append function and these lists are, by default, mutable. Python tuples are technically not lists, but we can treat them as immutable arrays. With Rust, we have **arrays** and **vectors**. Arrays are the most basic of the two. Defining and looping through an array is straightforward in Rust, as we can see in the following code:

```
let array: [i32; 3] = [1, 2, 3];

println!("array has {} elements", array.len());

for i in array.iter() {
```

```
        println!("{}", i);
}
```

If we try and append another integer onto our array with the push function, we will not be able to even if the array is mutable. If we add a fourth element to our array definition that is not an integer, the program will refuse to compile as all of the elements in the array have to be the same. However, this is not entirely true.

Later in this chapter, we will cover **structs**. In Python, the closest comparison to objects is structs as they have their own attributes and functions. Structs can also have **traits**, which we will also discuss later. In terms of Python, the closest comparison to **traits** is **mixins**. Therefore, a range of structs can be housed in an array if they all have the same trait in common. When looping through the array, the compiler will only allow us to execute functions from that trait as this is all we can ensure will be consistent throughout the array.

The same rules in terms of type or trait consistency also apply to **vectors**. However, vectors place their memory on the heap and are expandable. Like everything in Rust, they are, by default, immutable. However, applying the mut tag will enable us to add and manipulate the vector. In the following code, we define a vector, print the length of the vector, append another element to the vector, and then loop through the vector printing all elements:

```
let mut str_vector: Vec<&str> = vec!["one", "two", \
    "three"];

println!("{}", str_vector.len());

str_vector.push("four");

for i in str_vector.iter() {
    println!("{}", i);
}
```

This gives us the following output:

```
3
one
two
three
four
```

We can see that our append worked.

Considering the rules about consistency, vectors and arrays might seem a little restrictive to a Python developer. However, if they are, sit back and ask yourself why. Why would you want to put in a range of elements that do not have any consistency? Although Python allows you to do this, how could you loop through a list with inconsistent elements and confidently perform operations on them without crashing the program?

With this in mind, we are starting to see the benefits and safety behind this restrictive typing system. There are some ways in which we can put in different elements that are not structs bound by the same trait. Considering this, we will explore how we can store and access our varied data elements via hashmaps in Rust in the next section.

Replacing dictionaries with hashmaps

Hashmaps in Rust are essentially dictionaries in Python. However, unlike our previous vectors and arrays, we want to have a range of different data types housed in a hashmap (although we can also do this with vectors and arrays). To achieve this, we can use **Enums**. Enums are, well, Enums, and we have the exact same concept in Python. However, instead of it being an Enum, we merely have a Python object that inherits the Enum object as seen in the following code:

```python
from enum import Enum


class Animal(Enum):
    STRING = "string"
    INT = "int"
```

Here, we can use the Enum to save us from using **raw strings** in our Python code when picking a particular category. With a code editor known as an IDE, this is very useful, but it's understandable if a Python developer has never used them as they are not enforced anywhere. Not using them makes the code more prone to mistakes and harder to maintain when categories change and so on, but there is nothing in Python stopping the developer from just using a raw string to describe an option. In Rust, we are going to want our hashmap to accept strings and integers. To do this, we are going to have to carry out the following steps:

1. Create an Enum to handle multiple data types.

2. Create a new hashmap and insert values belonging to the Enum we created in *step 1*.

3. Test the data consistency by looping through the hashmap and match all possible outcomes.

4. Build a function that processes data extracted from the hashmap.

5. Use the function to process outcomes from getting a value from the hashmap.

Therefore, we are going to create an Enum that houses this using the following code:

```
enum Value {
    Str(&'static str),
    Int(i32),
}
```

Here, we can see that we have introduced the statement `'static`. This denotes a **lifetime** and basically states that the reference remains for the rest of the program's lifetime. We will cover lifetimes in the *Keeping track of scopes and lifetimes* section.

Now that we have defined our Enum, we can build our own **mutable hashmap** and insert an integer and a string into it with the following code:

```
use std::collections::HashMap;


let mut map = HashMap::new();


map.insert("one", Value::Str("1"));
map.insert("two", Value::Int(2));
```

Now that our hashmap is housing a single type that houses the two types we defined, we must handle them.

Remember, Rust has strong typing. Unlike Python, Rust will not allow us to compile unsafe code (Rust can compile in an unsafe context but this is not default behavior). We must handle every possible outcome, otherwise the compiler will refuse to compile. We can do this with a `match` statement as seen in the following code:

```
for (_key, value) in &map {

    match value {
        Value::Str(inside_value) => {
            println!("the following value is an str: {}", \
                inside_value);
        }
        Value::Int(inside_value) => {
            println!("the following value is an int: {}", \
```

```
                    inside_value);
        }
    }
}
```

In this code sample, we have looped through a borrowed reference to the hashmap using `&`. Again, we will cover borrowing later on in the *Understanding variable ownership* section. We prefix the `key` with a `_`. This is telling the compiler that we are not going to use the key. We don't have to do this as the compiler will still compile the code; however, it will complain by issuing a warning. The value that we are retrieving from the hashmap is our `Value` Enum. In this `match` statement, we can match the field of our Enum, and unwrap and access the inside value that we denote as `inside_value`, printing it to the console.

Running the code gives us the printout to the terminal as follows:

```
the following value is an int: 2
the following value is an str: 1
```

It must be noted that Rust is not going to let anything slip by the compiler. If we remove the match for our `Int` field for our Enum, then the compiler will throw the error seen here:

```
18 |              match value {
   |              ^^^^^ pattern '&Int(_)' not covered
   |
   = help: ensure that all possible cases are being
     handled,
   possibly by adding wildcards or more match arms
   = note: the matched value is of type '&Value'
```

This is because we have to handle every single possible outcome. Because we have been explicit that only values that can be housed in our Enum can be inserted into the hashmap, we know that there are only two possible types that can be extracted from our hashmap. We have nearly covered enough about hashmaps to use them effectively in Rust programs. One last concept that we must cover is the Enum called `Option`.

Considering that we have arrays and vectors, we will not be using our hashmaps primarily for looping through outcomes. Instead, we will be retrieving values from them when we need them. Like in Python, the hashmap has a get function. In Python, if the key that is being searched is not in the dictionary, then the get function will return None. It is then left to the developer to decide what to do with it. However, in Rust, the hashmap will return a Some or None. To demonstrate this, let's try to get a value belonging to a key that we know is not there:

1. Start by running the following code:

```
let outcome: Option<&Value> = map.get("test");
println!("outcome passed");
let another_outcome: &Value = \
    map.get("test").unwrap();
println!("another_outcome passed");
```

   Here, we can see that we can access the reference to the Value Enum wrapped in Option with the get function. We then directly access the reference to the Value Enum using the unwrap function.

2. However, we know that the test key is not in the hashmap. Because of this, the unwrap function will cause the program to crash, as seen in the following output from the previous code:

```
thread 'main' panicked at 'called 'Option::unwrap()'
on a 'None' value', src/main.rs:32:51
```

   We can see that the simple get function did not crash the program. However, we didn't manage to get the string "another_outcome passed" to print out to the console. We can handle this with a match statement.

   However, this is going to be a match statement within a match statement.

3. In order to reduce the complexity, we should explore Rust functions to process our value Enum. This can be done with the following code:

```
fn process_enum(value: &Value) -> () {
    match value {
        Value::Str(inside_value) => {
            println!("the following value is an str: \
            {}", inside_value);
        }
        Value::Int(inside_value) => {
```

```
            println!("the following value is an int: \
            {}", inside_value);
        }
    }
}
```

The function does not really give us any new logic to explore. The `-> ()` expression is merely stating that the function is not returning anything.

4. If we are going to return a string, for instance, the expression would be `->`
    `String`. We do not need the `-> ()` expression; however, it can be helpful
    for developers to quickly understand what's going on with the function. We can
    then use this function to process the outcome from our `get` function with the
    following code:

```
match map.get("test") {
    Some(inside_value) => {
        process_enum(inside_value);
    }
    None => {
        println!("there is no value");
    }
}
```

We now know enough to utilize hashmaps in our programs. However, we must notice that we have not really handled errors; we have either printed out that nothing was found or let the `unwrap` function just result in an error. Considering this, we will move on to the next section on handling errors in Rust.

Error handling in Rust

Handling errors in Python is straightforward. We have a `try` block that houses an `except` block underneath. In Rust, we have a `Result` wrapper. This works in the same way as an `Option`. However, instead of having `Some` or `None`, we have `Ok` or `Err`.

To demonstrate this, we can build on the hashmap that was defined in the previous section. We accept `Option` from a `get` function applied to the hashmap. Our function will check to see whether the integer retrieved from the hashmap is above a threshold. If it's above the threshold, we will return a true value. If not, then it is false.

The problem is that there might not be a value in Option. We also know that the Value Enum might not be an integer. If any of this is the case, we should return an error. If not, we return a Boolean. This function can be seen here:

```
fn check_int_above_threshold(threshold: i32,
        get_result: Option<&Value>) -> Result<bool, &'static \
        str> {
    match get_result {
        Some(inside_value) => {
            match inside_value {
                Value::Str(_) => return Err(
                    "str value was supplied as opposed to \
                    an int which is needed"),
                    Value::Int(int_value) => {
                        if int_value > &threshold {
                            return Ok(true)
                        }
                        return Ok(false)
                    }
                }
            }
        None => return Err("no value was supplied to be \
            checked")
    }
}
```

Here, we can see that the None result from Option instantly returns an error with a helpful message as to why we are returning an error. With the Some value, we utilize another match statement to return an error with a helpful message that we cannot supply a string to check the threshold if the Value is a string. It must be noted that Value::Str(_) has a _ in it. This means that we do not care what the value is because we are not going to use it. In the final part, we check to see whether the integer is above the threshold returning Ok values that are either true or false. We implement this function with the following code:

```
let result: Option<&Value> = map.get("two");
let above_threshold: bool = check_int_above_threshold(1, \
    result).unwrap();
```

```
println!("it is {} that the threshold is breached", \
    above_threshold);
```

This gives us the following output in the terminal:

```
it is true that the threshold is breached
```

If we up the first parameter in our check_int_above_threshold function to 3, we get the following output:

```
it is false that the threshold is breached
```

If we change the key in map.get to three, we get the following terminal output:

```
thread 'main' panicked at 'called 'Result::unwrap()'
on an 'Err' value: "no value was supplied to be checked"'
```

If we change the key in map.get to one, we get the following terminal output:

```
thread 'main' panicked at 'called 'Result::unwrap()' on
an 'Err' value: "str value was supplied as opposed to an
int
```

We can add extra signposting to the unwrap with the expect function. This function unwraps the result and adds an extra message to the printout if there is an error. With the following implementation, the message "an error happened" will be added to the error message:

```
let second_result: Option<&Value> = map.get("one");
let second_threshold: bool = check_int_above_threshold(1, \
    second_result).expect("an error happened");
```

We can also directly throw an error if needed with the following code:

```
panic!("throwing some error");
```

We can also check to see whether the result is an error by using the is_err function as seen here:

```
result.is_err()
```

This returns a `bool`, enabling us to alter the direction of our program if we come across an error. As we can see, Rust gives us a range of ways in which we can throw and manage errors.

We can now handle enough of Rust's quirks to write basic scripts. However, if the program gets a little more complicated, we fall into other pitfalls such as variable ownership and lifetimes. In the next section, we cover the basics of variable ownership so we can continue to use our variables throughout a range of functions and structs.

Understanding variable ownership

As we pointed out in the introduction discussing why we should use Rust, Rust doesn't have a garbage collector; however, it is still memory-safe. We do this to keep the resources low and the speed high. However, how do we achieve memory safety without a garbage collector? Rust achieves this by enforcing some strict rules around variable ownership.

Like typing, these rules are enforced when the code is being compiled. Any violation of these rules will stop the compilation process. This can lead to a lot of initial frustration for Python developers, as Python developers like to use their variables as and when they want. If they pass a variable into a function, they also expect that variable to still be able to be mutated outside the function if they want. This can lead to issues when implementing concurrent executions. Python also allows this by running expensive processes under the hood to enable the multiple references with cleanup mechanisms when the variable is no longer referenced.

As a result, this mismatch in coding style gives Rust the false label of having a steep learning curve. If we learn the rules, we only must rethink our code a little, as the helpful compiler enables us to adhere to them easily. You'll also be surprised how this approach is not as restrictive as it sounds. Rust's compile-time checking is done to protect against the following memory errors:

- **Use after frees**: This is where memory is accessed once it has been freed, which can cause crashes. It can also allow hackers to execute code via this memory address.

- **Dangling pointers**: This is where a reference points to a memory address that no longer houses the data that the pointer was referencing. Essentially, this pointer now points to null or random data.

- **Double frees**: This is where allocated memory is freed, and then freed again. This can cause the program to crash and increases the risk of sensitive data being revealed. This also enables a hacker to execute arbitrary code.

- **Segmentation faults**: This is where the program tries to access the memory it's not allowed to access.

- **Buffer overrun**: An example of this is reading off the end of an array. This can cause the program to crash.

Rust manages to protect against these errors by enforcing the following rules:

- Values are owned by the variables assigned to them.

- As soon as the variable goes out of scope, it is deallocated from the memory it is occupying.

- Values can be used by other variables, if we adhere to the conventions around copying, moving, immutable borrowing, and mutable borrowing.

To really feel comfortable navigating these rules in code, we will explore copying, moving, immutable borrowing, and mutable borrowing in more detail.

Copy

This is where the value is copied. Once it has been copied, the new variable owns the value, and the existing variable also owns its own value:

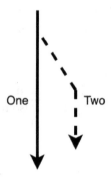

Figure 1.3 – Variable Copy path

As we can see with the pathway diagram in *Figure 1.3*, we can continue to use both variables. If the variable has a Copy trait, the variable will automatically copy the value. This can be achieved by the following code:

```
let one: i8 = 10;
let two: i8 = one + 5;
println!("{}", one);
println!("{}", two);
```

The fact that we can print out both the one and two variables means we know that one has been copied and the value of this copy has been utilized by two. Copy is the simplest reference operation; however, if the variable being copied does not have a Copy trait, then the variable must be moved. To understand this, we will now explore moving as a concept.

Move

This is where the value is moved from one variable to another. However, unlike Copy, the original variable no longer owns the value:

One

Two

Figure 1.4 – Variable Move path

Looking at the path diagram in *Figure 1.4*, we can see that one can no longer be used as it's been moved to two. We mentioned in the *Copy* section that if the variable does not have the Copy trait, then the variable is moved. In the following code, we show this by doing what we did in the *Copy* section but using String as this does not have a Copy trait:

```
let one: String = String::from("one");
let two: String = one + " two";
println!("{}", two);
println!("{}", one);
```

Running this gives the following error:

```
let one: String = String::from("one");
        --- move occurs because 'one' has type
        'String', which does not implement the
        'Copy' trait
let two: String = one + " two";
                  ------------ 'one' moved due to usage in operator
println!("{}", two);
```

```
println!("{}", one);
               ^^^ value borrowed here after move
```

This is really where the compiler shines. It tells us that the string does not implement the Copy trait. It then shows us where the move occurs. It is no surprise that many developers praise the Rust compiler. We can get round this by using the to_owned function with the following code:

```
let two: String = one.to_owned() + " two";
```

It is understandable to wonder why Strings do not have the Copy trait. This is because the string is a pointer to a string slice. Copying actually means copying bits. Considering this, if we were to copy strings, we would have multiple unconstrained pointers to the same string literal data, which would be dangerous. Scope also plays a role when it comes to moving variables. In order to see how scope forces movement, we need to explore immutable borrows in the next section.

Immutable borrow

This is where one variable can reference the value of another variable. If the variable that is borrowing the value falls out of scope, the value is not deallocated from memory as the variable borrowing the value does not have ownership:

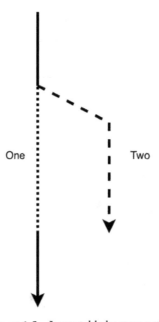

Figure 1.5 – Immutable borrow path

We can see with the path diagram in *Figure 1.5* that two borrows the value from one. When this is happening, one is kind of locked. We can still copy and borrow one; however, we cannot do a mutable borrow or move while two is still borrowing the value. This is because if we have mutable and immutable borrows of the same variable, the data of that variable could change through the mutable borrow causing an inconsistency. Considering this, we can see that we can have multiple immutable borrows at one time while only having one mutable borrow at any one time. Once two is finished, we can do anything we want to one again. To demonstrate this, we can go back to creating our own print function with the following code:

```
fn print(input_string: String) -> () {
    println!("{}", input_string);
}
```

With this, we create a string and pass it through our print function. We then try and print the string again, as seen in the following code:

```
let one: String = String::from("one");
print(one);
println!("{}", one);
```

If we try and run this, we will get an error stating that one was moved into our print function and therefore cannot be used in println!. We can solve this by merely accepting a borrow of a string using & in our function, as denoted in the following code:

```
fn print(input_string: &String) -> () {
    println!("{}", input_string);
}
```

Now we can pass a borrowed reference into our print function. After this, we can still access the | variable, as seen in the following code:

```
let one: String = String::from("one");
print(&one);
let two: String = one + " two";
println!("{}", two);
```

Borrows are safe and useful. As our programs grow, immutable borrows are safe ways to pass variables through to other functions in other files. We are nearly at the end of our journey toward understanding the rules. The only concept left that we must explore is mutable borrows.

Mutable borrow

This is where another variable can reference and write the value of another variable. If the variable that is borrowing the value falls out of scope, the value is not deallocated from memory as the variable borrowing the value does not have ownership. Essentially, a mutable borrow has the same path as an immutable borrow. The only difference is that while the value is being borrowed, the original variable cannot be used at all. It will be completely locked down as the value might be altered when being borrowed. The mutable borrow can be moved into another scope like a function, but cannot be copied as we cannot have multiple mutable references, as stated in the previous section.

Considering all that we have covered on borrowing, we can see a certain theme. We can see that scopes play a big role in implementing the rules that we have covered. If the concept of scopes is unclear, passing a variable into a function is changing scope as a function is its own scope. To fully appreciate this, we need to move on to exploring scopes and lifetimes.

Keeping track of scopes and lifetimes

In Python, we do have the concept of **scope**. It is generally enforced in functions. For instance, we can call the Python function defined here:

```python
def add_and_square(one: int, two: int) -> int:
    total: int = one + two
    return total * total
```

In this case, we can access the return variable. However, we will not be able to access the `total` variable. Outside of this, most of the variables are accessible throughout the program. With Rust, it is different. Like typing, Rust is aggressive with scopes. Once a variable is passed into a scope, it is deleted when the scope is finished. Rust manages to maintain memory safety without garbage collection with the borrowing rules. Rust deletes its variables without garbage collection by wiping all variables out of scope. It can also define scopes with curly brackets. A classic way of demonstrating scopes can be done by the following code:

```rust
fn main() {
    let one: String = String::from("one");

    // start of the inner-scope
    {
        println!("{}", &one);
```

```
        let two: String = String::from("two");
    }
    // end of the inner-scope

    println!("{}", one);
    println!("{}", two);
}
```

If we try and run this code, we get the error code defined here:

```
println!("{}", two);
               ^^^ not found in this scope
```

We can see that the variable one can be accessed in the inner-scope as it was defined outside the outer-scope. However, the variable two is defined in the inner-scope. Once the inner-scope is finished, we can see by the error that we cannot access the variable two outside the inner-scope. We must remember that the scope of functions is a little stronger. From revising borrowing rules, we know that when we move a variable into the scope of a function, it cannot be accessed outside of the scope of the function if the variable is not borrowed as it is moved. However, we can still alter a variable inside another scope like another function, and still then access the changed variable. To do this, we must do a mutable borrow, and then must dereference (using *) the borrowed mutable variable, alter the variable, and then access the altered variable outside the function, as we can see with the following code:

```
fn alter_number(number: &mut i8) {
    *number += 1
}
fn print_number(number: i8) {
    println!("print function scope: {}", number);
}

fn main() {
    let mut one: i8 = 1;
    print_number(one);
    alter_number(&mut one);
    println!("main scope: {}", one);
}
```

This gives us the following output:

```
print function scope: 1
main scope: 2
```

With this, we can see that that if we are comfortable with our borrowing, we can be flexible and safe with our variables. Now that we have explored the concept of scopes, this leads naturally to lifetimes, as lifetimes can be defined by scopes. Remember that a borrow is not sole ownership. Because of this, there is a risk that we could reference a variable that's deleted. This can be demonstrated in the following classic demonstration of a lifetime:

```
fn main() {
    let one;
    {
        let two: i8 = 2;
        one = &two;
    } // ----------------------> two lifetime stops here
    println!("r: {}", one);
}
```

Running this code gives us the following error:

```
    one = &two;
         ^^^^ borrowed value does not live long enough
} // ----------------------> two lifetime stops here
- 'two' dropped here while still borrowed
println!("r: {}", one);
                  --- borrow later used here
```

What has happened here is that we state that there is a variable called one. We then define an inner-scope. Inside this scope, we define an integer two. We then assign one to be a reference of two. When we try and print one in the outer-scope, we can't, as the variable it is pointing to has been deleted. Therefore, we no longer get the issue that the variable is out of scope, it's that the lifetime of the value that the variable is pointing to is no longer available, as it's been deleted. The lifetime of two is shorter than the lifetime of one.

While it is great that this is flagged when compiling, Rust does not stop here. This concept also translates functions. Let's say that we build a function that references two integers, compares them, and returns the highest integer reference. The function is an isolated piece of code. In this function, we can denote the lifetimes of the two integers. This is done by using the ' prefix, which is a lifetime notation. The names of the notations can be anything you wish, but it's a general convention to use a, b, c, and so on. Let's look at an example:

```
fn get_highest<'a>(first_number: &'a i8, second_number: &'\
   a       i8) -> &'a i8 {
      if first_number > second_number {
            return first_number
      } else {
            return second_number
      }
}
fn main() {
      let one: i8 = 1;
      {
            let two: i8 = 2;
            let outcome: &i8 = get_highest(&one, &two);
            println!("{}", outcome);
      }
}
```

As we can see, the first_number and second_number variables have the same lifetime notation of a. This means that they have the same lifetimes. We also have to note that the get_highest function returns an i8 with a lifetime of a. As a result, both first_number and second_number variables can be returned, which means that we cannot use the outcome variable outside of the inner-scope. However, we know that our lifetimes between the variables one and two are different. If we want to utilize the outcome variable outside of the inner-scope, we must tell the function that there are two different lifetimes. We can see the definition and implementation here:

```
fn get_highest<'a, 'b>(first_number: &'a i8, second_ \
   number:    &'b i8) -> &'a i8 {
      if first_number > second_number {
            return first_number
      } else {
```

```
        return &0
    }
}
fn main() {
    let one: i8 = 1;
    let outcome: &i8;
    {
        let two: i8 = 2;
        outcome = get_highest(&one, &two);
    }
    println!("{}", outcome);
}
```

Again, the lifetime a is returned. Therefore, the parameter with the lifetime b can be defined in the inner-scope as we are not returning it in the function. Considering this, we can see that lifetimes are not exactly essential. We can write comprehensive programs without touching lifetimes. However, they are an extra tool. We don't have to let scopes completely constrain us with lifetimes.

We are now at the final stages of knowing enough Rust to be productive Rust developers. All we need to understand now is building structs and managing them with macros. Once this is done, we can move onto the next chapter of structuring Rust programs. In the next section, we will cover the building of structs.

Building structs instead of objects

In Python, we use a lot of objects. In fact, everything you work with in Python is an object. In Rust, the closest thing we can get to objects is structs. To demonstrate this, let's build an object in Python, and then replicate the behavior in Rust. For our example, we will build a basic stock object as seen in the following code:

```
class Stock:

    def __init__(self, name: str, open_price: float,\
        stop_loss: float = 0.0, take_profit: float = 0.0) \
            -> None:
        self.name: str = name
        self.open_price: float = open_price
        self.stop_loss: float = stop_loss
```

```
        self.take_profit: float = take_profit
        self.current_price: float = open_price

    def update_price(self, new_price: float) -> None:
        self.current_price = new_price
```

Here, we can see that we have two mandatory fields, which are the name and price of the stock. We can also have an optional stop loss and an optional take profit. This means that if the stock crosses one of these thresholds, it forces a sale, so we don't continue to lose more money or keep letting the stock rise to the point where it crashes. We then have a function that merely updates the current price of the stock. We could add extra logic here on the thresholds for it to return a bool for whether the stock should be sold or not if needed. For Rust, we define the fields with the following code:

```
struct Stock {
    name: String,
    open_price: f32,
    stop_loss: f32,
    take_profit: f32,
    current_price: f32
}
```

Now we have our fields for the struct, we need to build the constructor. We can build functions that belong to our struct with an `impl` block. We build our constructor with the following code:

```
impl Stock {

    fn new(stock_name: &str, price: f32) -> Stock {
        return Stock{
            name: String::from(stock_name),
            open_price: price,
            stop_loss: 0.0,
            take_profit: 0.0,
            current_price: price
        }
    }
}
```

Here, we can see that we have defined some default values for some of the attributes. To build an instance, we use the following code:

```
let stock: Stock = Stock::new("MonolithAi", 95.0);
```

However, we have not exactly replicated our Python object. In the Python object __ init__, there were some optional parameters. We can do this by adding the following functions to our impl block:

```
fn with_stop_loss(mut self, value: f32) -> Stock {
    self.stop_loss = value;
    return self
}
fn with_take_profit(mut self, value: f32) -> Stock {
    self.take_profit = value;
    return self
}
```

What we do here is take in our struct, mutate the field, and then return it. Building a new stock with a stop loss involves calling our constructor followed by the with_stop_loss function as seen here:

```
let stock_two: Stock = Stock::new("RIMES",\
    150.4).with_stop_loss(55.0);
```

With this, our RIMES stock will have an open price of 150.4, current price of 150.4, and a stop loss of 55.0. We can chain multiple functions as they return the stock struct. We can create a stock struct with a stop loss and a take profit with the following code:

```
let stock_three: Stock = Stock::new("BUMPER (former known \
   as ASF)", 120.0).with_take_profit(100.0).\
    with_stop_loss(50.0);
```

We can continue chaining with as many optional variables as we want. This also enables us to encapsulate the logic behind defining these attributes. Now that we have all our constructor needs sorted, we need to edit the update_price attribute. This can be done by implementing the following function in the impl block:

```
fn update_price(&mut self, value: f32) {
    self.current_price = value;
}
```

This can be implemented with the following code:

```
let mut stock: Stock = Stock::new("MonolithAi", 95.0);
stock.update_price(128.4);
println!("here is the stock: {}", stock.current_price);
```

It has to be noted that the stock needs to be mutable. The preceding code gives us the following printout:

```
here is the stock: 128.4
```

There is only one concept left to explore for structs and this is traits. As we have stated before, traits are like Python mixins. However, **traits** can also act as a data type because we know that a struct that has the trait has those functions housed in the trait. To demonstrate this, we can create a `CanTransfer` trait in the following code:

```
trait CanTransfer {
    fn transfer_stock(&self) -> ();

    fn print(&self) -> () {
        println!("a transfer is happening");
    }
}
```

If we implement the trait for a struct, the instance of the struct can utilize the `print` function. However, the `transfer_stock` function doesn't have a body. This means that we must define our own function if it has the same return value. We can implement the trait for our struct using the following code:

```
impl CanTransfer for Stock {
    fn transfer_stock(&self) -> () {
        println!("the stock {} is being transferred for \
            £{}", self.name, self.current_price);
    }
}
```

We can now use our trait with the following code:

```
let stock: Stock = Stock::new("MonolithAi", 95.0);
stock.print();
stock.transfer_stock();
```

This gives us the following output:

```
a transfer is happening
the stock MonolithAi is being transferred for £95
```

We can make our own function that will print and transfer the stock. It will accept all structs that implement our `CanTransfer` trait and we can use all the trait's functions in it, as seen here:

```
fn process_transfer(stock: impl CanTransfer) -> () {
    stock.print();
    stock.transfer_stock();
}
```

We can see that traits are a powerful alternative to object inheritance; they reduce the amount of repeated code for structs that fit in the same group. There is no limit to the number of traits that a struct can implement. This enables us to plug traits in and out, adding a lot of flexibility to our structs when maintaining code.

Traits are not the only way by which we can manage how structs interact with the rest of the program; we can achieve metaprogramming with macros, which we will explore in the next section.

Metaprogramming with macros instead of decorators

Metaprogramming can generally be described as a way in which the program can manipulate itself based on certain instructions. Considering the strong typing Rust has, one of the simplest ways that we can metaprogram is by using generics. A classic example of demonstrating generics is through coordinates:

```
struct Coordinate <T> {
        x: T,
        y: T
    }
fn main() {
    let one = Coordinate{x: 50, y: 50};
    let two = Coordinate{x: 500, y: 500};
    let three = Coordinate{x: 5.6, y: 5.6};
}
```

What is happening here is that the compiler is looking through all the uses of our struct throughout the whole program. It then creates structs that have those types. Generics are a good way of saving time and getting the compiler to write repetitive code. While this is the simplest form of metaprogramming, another form of metaprogramming in Rust is **macros**.

You may have noticed throughout the chapter that some of the functions that we use, such as the `println!` function, have an `!` at the end. This is because it is not technically a function, it is a macro. The `!` denotes that the macro is being called. Defining our own macros is a blend of defining our own function and using lifetime notation within a `match` statement within the function. To demonstrate this, we can define our own macro that capitalizes the first character in a string passed through it with the following code:

```
macro_rules! capitalize {
        ($a: expr) => {
            let mut v: Vec<char> = $a.chars().collect();
            v[0] = v[0].to_uppercase().nth(0).unwrap();
            $a = v.into_iter().collect();
        }
    }
fn main() {
    let mut x = String::from("test");
    capitalize!(x);
    println!("{}", x);
}
```

Instead of using the `fn` term that is used for defining functions, we define our macro using `macro_rules!`. We then say that the `$a` is the expression passed into the macro. We then get the expression, convert it into a vector of chars, uppercase the first character, and then convert it back to a string. It must be noted that the macro that we defined does not return anything, and we do not assign any variable when calling our macro in the main function. However, when we print the x variable at the end of the main function, it is capitalized. Therefore, we can deduce that our macro is altering the state of the variable.

However, we must remember that macros are a last resort. Our example shows that our macro alters the state even though it is not directly demonstrated in the `main` function. As the complexity of the program grows, we could end up with a lot of brittle, highly coupled processes that we are not aware of. If we change one thing, it could break five other things. For capitalizing the first letter, it is better to just build a function that does this and returns a string value.

Macros do not just stop at what we have covered, they also have the same effect as our **decorators** in Python. To demonstrate this, let's look at our coordinate again. We can generate our coordinate and then pass it through a function so it can be moved. We then try to print the coordinate outside of the function with the following code:

```
struct Coordinate {
    x: i8,
    y: i8
}
fn print(point: Coordinate) {
    println!("{} {}", point.x, point.y);
}
fn main() {
    let test = Coordinate{x: 1, y:2};
    print(test);
    println!("{}", test.x)
}
```

It will be expected that Rust will refuse to compile the code because the coordinate has been moved into the scope of the `print` function that we created and therefore we cannot use it in the final `println!`. We could borrow the coordinate and pass that through to the function. However, there is another way we can do this. Remember that integers passed through functions without any trouble because they had a `Copy` trait. Now, we could try and code a `Copy` trait ourselves, but this would be convoluted and would require advanced knowledge. Luckily for us, we can implement the `Copy` and `Clone` traits by utilizing a `derive` macro with the following code:

```
#[derive(Clone, Copy)]
struct Coordinate {
    x: i8,
    y: i8
}
```

With this, our code works as we copy the coordinate when passing it through the function. Macros can be utilized by many packages and frameworks, from **JavaScript Object Notation (JSON)** serialization to entire web frameworks. In fact, here is the classic example of running a basic server in the Rocket framework:

```
#![feature(proc_macro_hygiene, decl_macro)]


#[macro_use] extern crate rocket;


#[get("/hello/<name>/<age>")]
fn hello(name: String, age: u8) -> String {
    format!("Hello, {} year old named {}!", age, name)
}
fn main() {
    rocket::ignite().mount("/", routes![hello]).launch();
}
```

This is a striking resemblance to the Python Flask application example at the beginning of the chapter. These macros are acting exactly like our decorators in Python, which is not surprising as a decorator in Python is a form of metaprogramming that wraps a function.

This wraps up our brief introduction to the Rust language for Python developers. We are now able to move on to other concepts, such as structuring our code and building fully fledged programs coded in Rust.

Summary

In this chapter, we explored the role of Rust in today's landscape, showing that Rust's paradigm-changing position is a result of being memory-safe, while not having any garbage collection. With this, we understood why it beats most languages (including Golang) when it comes to speed. We then went over the quirks that Rust has when it comes to strings, lifetimes, memory management, and typing, so we can write safe and efficient Rust code as Python developers. We then covered structs and traits to the point where we could mimic the basic functionality of a Python object with mixins, utilizing their traits as types for the Rust struct while we were at it.

We covered the basic concepts of lifetimes and borrowing. This enables us to have more control over how we implement our structs and functions within our program, giving us multiple avenues to turn to when solving a problem. With all this, we can safely code single-page applications with confidence over concepts that would stump someone who has never coded in Rust. However, we know, as experienced Python developers, that any serious program worth coding spans multiple pages. Considering this, we can use what we have learned here to move on to the next chapter, where we set up a Rust environment on our own computers and learn how to structure Rust code over multiple files, enabling us to get one step closer to building packages in Rust and installing them with `pip`.

Questions

1. Why can we not simply copy a String?

2. Rust has strong typing. In which two ways can we enable a container such as a vector or hashmap to contain multiple different types?

3. How are Python decorators and Rust macros the same?

4. What is the Python equivalent to a `main` function in Rust?

5. Why can we get a higher integer value with the same number of bytes with an unsigned integer than a signed integer?

6. Why do we have to be strict with lifetimes and scopes when coding in Rust?

7. Can we reference a variable when it has been moved?

8. What can you do to an original variable if it is currently being borrowed in an immutable state?

9. What can you do to an original variable if it is currently being borrowed in a mutable state?

Answers

1. This is because a String is essentially a pointer to `Vec<u8>` with some metadata. If we copy this, then we will have multiple unconstrained pointers to the same string literal, which will introduce errors with concurrency, mutability, and lifetimes.

2. We can use an Enum, which means that the type being accepted into the container can be one of those types housed in the Enum. When reading the data, we can then use a `match` statement to manage all possible data types that could be read from the container. The second way is to create a trait that multiple different structs implement. However, the only interaction that we can have from the container read when this is the case is the functions that the trait implements.

3. They both wrap around the code and alter the implementation or attributes of the code that they are wrapping without directly returning anything.

4. The Python equivalent is `if __name__ == "__main__":`.

5. A signed integer must accommodate positive and negative values, whereas an unsigned integer only accommodates positive values.

6. This is because there is no garbage collection; as a result, variables get deleted when they shift out of the scope of where they were created. If we do not consider lifetimes, we could reference a variable that has been deleted.

7. No, the ownership of the variable has essentially been moved and there are no references to the original variable anymore.

8. We can still copy and borrow the original variable; however, we cannot perform a mutable borrow.

9. We cannot use the original variable at all as the state of the variable might be altered.

Further reading

- *Hands-On Functional Programming in Rust* (2018) by Andrew Johnson, Packt Publishing

- *Mastering Rust* (2019) by Rahul Sharma and Vesa Kaihlavirta, Packt Publishing

- *The Rust Programming Language* (2018): `https://doc.rust-lang.org/stable/book/`

2
Structuring Code in Rust

Now that we have gotten to grips with the basics of Rust, we can move on to structuring code over several files so we can actually solve problems with Rust. In order to do this, we will have to understand how to manage dependencies as well as how to compile a basic and structured application. We also have to consider the isolation of code so we can reuse it and keep the development of the application agile, enabling us to make changes quickly without much pain. After covering this, we will also get the application to interact with the user directly by accepting user commands. We will also utilize Rust crates. A crate is a binary or library that we import and use.

In this chapter, we will cover the following topics:

- Managing our code with crates and Cargo instead of `pip`
- Structuring code over multiple files and modules
- Building module interfaces
- Interacting with the environment

Technical requirements

We are no longer going to be implementing simple single-page applications that do not rely on any third-party dependencies as we did in the first chapter. As a result, you will have to directly install Rust onto your computer. We will also be managing third-party dependencies through Cargo. You can install Rust and Cargo on your computer here: `https://www.rust-lang.org/tools/install`.

At the time of writing this, the best **integrated development environment (IDE)** by far for writing Rust is Visual Studio Code. It has a range of Rust plugins that can help you keep track of and check your Rust code. It can be installed using this link: `https://code.visualstudio.com/download`.

You can find all the code files in the GitHub repository for this chapter: `https://github.com/PacktPublishing/Speed-up-your-Python-with-Rust/tree/main/chapter_two`.

Managing our code with crates and Cargo instead of pip

Building our own application is going to involve the following steps:

1. Create a simple Rust file and run it.
2. Create a simple application using Cargo.
3. Run our application using Cargo.
4. Manage dependencies with Cargo.
5. Use a third-party crate to serialize JSON.
6. Document our application with Cargo.

Before we start structuring our program with Cargo, we should compile a basic Rust script and run it:

1. To do this, make a file called `hello_world.rs` with the main function housing the `println!` function with a string, as we can see here:

```
fn main() {
    println!("hello world");
}
```

2. Once this is done, we can navigate to the file and run the `rustc` command:

```
rustc hello_world.rs
```

3. This command compiles the file into a binary to be run. If we compile on Windows, we can run the binary with the following command:

```
.\hello_world.exe
```

4. If we compile it on Linux or Mac, we can run it with the following command:

```
./hello_world
```

The console should then print out the string. While this can come in useful when building a standalone file, it is not recommended for managing programs spanning multiple files. It is not even recommended when relying on dependencies. This is where Cargo comes in. Cargo manages everything – the running, testing, documentation, building, and dependency out of the box – with a few simple commands.

Now that we have a basic understanding of how to compile a basic file, we can move on to building a fully fledged application:

1. In your terminal, navigate to where you want your application to sit, and create a new project called `wealth_manager` as follows:

```
cargo new wealth_manager
```

This will create our application with the following structure:

```
└── wealth_manage
        ├── .git
    ├── Cargo.toml
    ├── .gitignore
    └── src
            └── main.rs
```

Here, we can see that Cargo has built the basic structure of an application that can manage compilations, work with GitHub, and manage dependencies right out of the box. The metadata and dependencies of our application are defined in the `Cargo.toml` file.

In order to perform Cargo commands on this application, our terminal is going to have to be in the same directory as the `Cargo.toml` file. The code that we are going to be altering that makes up our application is housed in the `src` directory. Our entry point for the whole application is in the `main.rs` file. In Python, we can have multiple entry points, and we will explore these in *Chapter 4, Building pip Modules in Python*, where we will build pure Python packages for the first time. If we open the `.gitignore` file, we should have the following:

```
/target
```

This is not a mistake; this is how clean Rust is. Everything that Cargo produces when it comes to compiling, documenting, caching, and so on is all stored in the target directory.

2. Right now, all we have is the main file that has a printout to the console saying *"hello world."* We can run this with the following command:

```
cargo run
```

3. With this command, we get the following output in the terminal:

```
Compiling wealth_manager v0.1.0 (/Users/maxwellflitton
/Documents/github/book_two/chapter_two/wealth_manager)

Finished dev [unoptimized + debuginfo] target(s) in 0.45s

Running 'target/debug/wealth_manager'

Hello, world!
```

Here, we can see that the application is being compiled. Once it's been compiled, it states that the compilation process has finished.

However, we have to note that it states that the finished process is unoptimized with debug info. This means that the compiled product is not as fast as it could be; however, it does contain debugging info if needed. This type of compilation is fast compared to an optimized version and is to be used when developing an application, as opposed to live production environments. We then see that the binary file that has been compiled is in `target/debug/wealth_manager`, and this is then run, resulting in the `hello_world.rs` output.

4. If we want to run a release, we simply run the following command:

```
cargo run --release
```

This compiles an optimized version of our app in the `./target/release/` directory under the binary `wealth_manager`. If we just want to compile our application without running it, we can simply swap the `run` command for `build`.

Now that we have got our application running, let's explore how we manage the metadata around it. This can all be done by editing the `Cargo.toml` file. When we open this, we get the following:

```
[package]
name = "wealth_manager"
version = "0.1.0"
authors = ["maxwellflitton"]
edition = "2018"


[dependencies]
```

The name, version, and authors are fairly straightforward. Here are the effects each section has on the project:

*   If we change the `name` value in the `Cargo.toml` file, then new binaries with that name will be made when we build or run our application. The old ones will still be there too.

*   `version` is for distribution on services such as `crates.io` if we want to open source our application for others to use. The authors are required for this as well, and our application will still compile and run locally if it's not there.

*   `edition` is the edition of Rust that we are using. Rust gets updated frequently. These updates accumulate through time, and every two to three years, the smoothed-out new features are packaged, documented, and added to a new edition. The latest edition (2021) is available at `https://devclass.com/2021/10/27/rust-1-56-0-arrives-delivering-rust-2021-edition-support/`.

*   We also have `dependencies`. This is where we can import third-party crates.

To see how this works, let's use a crate to convert a data structure of stock into JSON and then print it. Writing the code ourselves would be a bit of a headache. Luckily, we can install the serde crate and use the json! macro. In order for Cargo to download and install the crate, we fill our dependencies section in our Cargo.toml file with the code given here:

```
[dependencies]
serde="1.0.117"
serde_json="1.0.59"
```

In our main.rs file, we then import the macro and struct needed to convert data about a stock into JSON and then print it out in the following code:

```rust
use serde_json::{json, Value};

fn main() {
    let stock: Value = json!({
        "name": "MonolithAi",
        "price": 43.7,
        "history": [19.4, 26.9, 32.5]
    });

    println!("first price: {}", stock["history"][0]);
    println!("{}", stock.to_string());
}
```

It is important to note that we are returning a Value struct from the serde_json value. In order to see how we can use the return value, we can explore the documentation of the struct. This is when we get to see that Rust's documentation system is very comprehensive. We can find the documentation of the struct here: https://docs.rs/serde_json/1.0.64/serde_json/enum.Value.html.

We can see in *Figure 2.1* that the documentation covers all of the functions that the struct supports. Our json! macro is returning Object(Map<String, Value>). We also have a range of other values, depending on how we call the json! macro. The documentation also covers a range of functions that we can exploit to check what type the value is, whether the JSON value is null, and ways in which we can cast the JSON value as a particular type:

Figure 2.1 – Documentation of the serde_json value

When we perform a Cargo `run` command, we will see Cargo compiling the crates that we defined in the dependencies. We then see the compilation of our own app, and the printout of the price and the data related to the stock, as shown here:

```
first price: 19.4
{"history":[19.4,26.9,32.5], "name":"MonolithAi",\
   "price":43.7}
```

Going back to the documentation, we can create our own. This is straightforward; we do not have to install anything. All we have to do is create documentation in the code, like **docstrings** in Python. In order to demonstrate this, we can create a function that adds two variables together and defines the docstring, as seen in the following code:

```
/// Adds two numbers together.
///
/// # Arguments
/// * one (i32): one of the numbers to be added
/// * two (i32): one of the numbers to be added
///
/// # Returns
/// (i32): the sum of param one and param two
///
/// # Usage
/// The function can be used by the following code:
///
/// '''rust
/// result: i32 = add_numbers(2, 5);
/// '''
fn add_numbers(one: i32, two: i32) -> i32 {
    return one + two
}
```

We can see that this documentation is Markdown! This example is overkill for this type of function. A standard developer should be able to implement this function without any examples. For more complex functions and structs, it is worth noting that there is nothing stopping us from documenting code examples on how to implement what we are documenting. Building the documentation only requires the command here:

```
cargo doc
```

After the process has finished, we can open the documentation with the following command:

```
cargo doc --open
```

This opens up the documentation in a web browser, as shown in *Figure 2.2*:

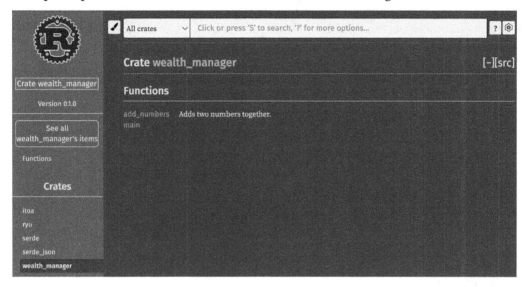

Figure 2.2 – Documentation view of our module

What we can see here is that our `main` and `add_numbers` functions are available. We can also see on the left that the dependencies that were installed are also available. If we click on our `add_numbers` function, we get to see the Markdown that we wrote, as shown in *Figure 2.3*:

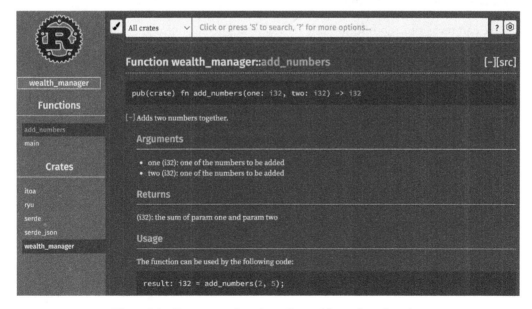

Figure 2.3 – Documentation view of our add_numbers function

Here we have it – we can create interactive documentation of our code as we build our application. It has to be noted that the rest of the book will not have Markdown in the code snippets; otherwise, this would simply extend the book to an unnecessary length. However, it is good practice to document all structs and functions as you code.

Now that we have run our code, set up a basic application structure, and documented our code, we are ready to move on to the next section of structuring our application over multiple files.

Structuring code over multiple files and modules

In order to build our module, we are going to carry out the following steps:

1. Map out our file and folder structure.
2. Create our Stock structs.
3. Link our Stock struct to the main file.
4. Use our stocks module in the main file.
5. Add code from another module.

Now that we are at the stage of building out our application over multiple files, we have to define our first module in our application, which is the **stocks module**:

1. We can make our module have the structure defined as follows:

    ```
    ├── main.rs
    └── stocks
        ├── mod.rs
        └── structs
            ├── mod.rs
            └── stock.rs
    ```

    We have taken this structure to enable flexibility; if we need to add more structs, we can do so in the structs directory. We can also add other directories alongside the structs directory. For instance, we might want to build a mechanism for storing the data for our stocks. This can be achieved by adding a storage directory in the stocks directory and using this throughout the module as and when it is needed.

2. For now, we simply want to create a stock struct in our `stocks` module, import it into our `main.rs` file, and use it. Our first step is to define our `Stock` struct in our `stock.rs` file with this code:

```
pub struct Stock {
    pub name: String,
    pub open_price: f32,
    pub stop_loss: f32,
    pub take_profit: f32,
    pub current_price: f32
}
```

This looks familiar, as it is the same as the `Stock` struct that we defined in the previous chapter. However, there is a slight difference. We must note that there is a `pub` keyword before the struct definition and each field definition. This is because we have to declare them public before we can use them outside the file. This also applies to functions implemented in the same file, as shown in the following code:

```
impl Stock {
    pub fn new(stock_name: &str, price: f32) -> \
      Stock {
        return Stock{
            name: String::from(stock_name),
            open_price: price,
            stop_loss: 0.0,
            take_profit: 0.0,
            current_price: price
        }
    }
    pub fn with_stop_loss(mut self, value: f32) \
      -> Stock {
        self.stop_loss = value;
        return self
    }
    pub fn with_take_profit(mut self, value: f32) \
      -> Stock {
        self.take_profit = value;
        return self
    }
```

```
        pub fn update_price(&mut self, value: f32) {
            self.current_price = value;
        }
    }
```

We can see that we now have a public struct that is available with all its functions.

We now have to enable our struct to be used in the main.rs file. This is where the mod.rs files come in. mod.rs files are essentially __init__.py files in Python. They show that the directory is a module. However, unlike Python, Rust data structures need to be publicly declared in order to be accessed from other files. We can see how the struct is passed through our stocks module to our main.rs file in *Figure 2.4*:

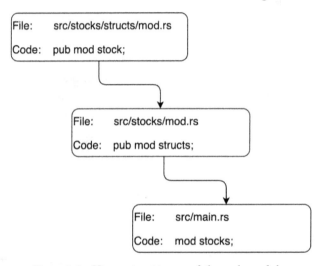

Figure 2.4 – How a struct is passed through modules

Here, we can see that we are merely publicly declaring the struct in the module furthest away from main.rs in the mod.rs file belonging to that directory. We then publicly declare the structs module in the stocks mod.rs file. Here is a good time to explore the mod expression that declares modules. If we want to, we can declare multiple modules in a single file. It must be stressed that this is not happening in our example. We could declare module one and module two in a single file with the following code:

```
mod one {
    . . .
}
Mod two {
    . . .
}
```

Now that we have defined our modules in our main example project, we just declare the stocks module in the main.rs file. The reason why this is not a public declaration is that the main.rs file is the entry point of our application; we will not be importing this module into anything else:

1. Now that our struct is available, we can simply use it as we would if it was defined in the same file with the following:

```
mod stocks;
use stocks::structs::stock::Stock;

fn main() {
    let stock: Stock = Stock::new("MonolithAi", 36.5);
    println!("here is the stock name: {}",\
        stock.name);
    println!("here is the stock name: {}",\
        stock.current_price);
}
```

2. Running this code unsurprisingly gives us this:

```
here is the stock name: MonolithAi
here is the stock name: 36.5
```

Now that we have the basics of using structs from different files covered, we can move on to exploring other pathways of accessing data structures from other files in order to be more flexible:

1. The first concept we have to explore is accessing from files in the same directory. In order to demonstrate this, we can do a throwaway example of building a print function in the structs. In a new file with the src/stocks/structs/utils.rs path, we can create a toy function that merely prints out that the constructor for the struct is firing, as shown in the following code:

```
pub fn constructor_shout(stock_name: &str) -> () {
    println!("the constructor for the {} is firing", \
        stock_name);
}
```

2. We then declare it in our `src/stocks/structs/mod.rs` file with the following code:

```
pub mod stock;
mod utils;
```

It must be noted that we are not making it public; we just declare it instead. Nothing is stopping us from making it public; however, with the non-public approach, we only allow files within the directory of `src/stocks/structs/` to access it.

3. We now want our `Stock` struct to access it and use it in our constructor, which can be done with an import in `src/stocks/structs/stock.rs` with the following line:

```
use super::utils::constructor_shout;
```

4. If we want to move our reference to the `src/stocks/` directory, we can use `super::super`. We can chain as many supers as we want, depending on how deep the tree is. It has to be noted that we can only access what is declared in the `mod.rs` file of the directory. In our `src/stocks/structs/stock.rs` file, we can now use the function in our constructor with the following code:

```
pub fn new(stock_name: &str, price: f32) -> Stock {
    constructor_shout(stock_name);
    return Stock{
        name: String::from(stock_name),
        open_price: price,
        stop_loss: 0.0,
        take_profit: 0.0,
        current_price: price
    }
}
```

5. Now, if we run our application, we will get the following printout in the terminal:

```
the constructor for the MonolithAi is firing
here is the stock name: MonolithAi
here is the stock name: 36.5
```

We can see that the program runs in exactly the same way, with the additional line from the `util` function that we imported from. If we create another module, we can access our `stocks` module from it, because the `stocks` module is defined in the `main.rs` file.

While we have managed to access data structures from different files and modules, this is not very scalable, and there are going to be some rules in which we implement stocks. In order to enable us to write scalable safe code, we need to lock down the functionality with interfaces in the next section.

Building module interfaces

Unlike Python, where we can import anything we want from anywhere and at the most our IDE will just give us a syntax highlight, Rust will actively not compile if we try and access data structures that have not explicitly been made public. This gives us an opportunity to really lock down our modules and enforce functionality through an interface.

However, before we get started with this, let's fully explore what functionality we will be locking down. It is good practice to keep modules as isolated as possible. In our `stocks` module, the logic should only be around how to handle stocks and nothing else. This might seem a little overkill, but when we think about it, we quickly realize that this module is going to scale when it comes to complexity.

For the demonstrative purposes of this chapter, let's just build the functionality for a stock order. We can either buy or sell a stock. These stock orders come in multiples. It's fairly common to buy *n* stocks of a company. We will also have to check to see whether the stock order is short or long. With a short order, we borrow money from the broker, buy stocks with that money, sell them instantly, and then buy the stocks back at a later date. If the stock price goes down, we make money, as we keep the difference when repaying to the broker. If we go long, we just buy the stock and hold it. If it goes up, we make money, so depending on the order, there will be different outcomes.

We have to remember that this is not a book for developing software around stock markets, so we need to keep the details simple to avoid losing ourselves. A simple approach for us to take to demonstrate interfaces is to take a layered approach, as described in *Figure 2.5*:

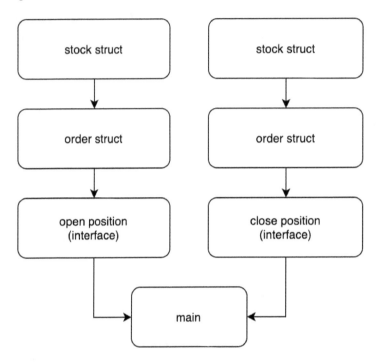

Figure 2.5 – Approach to a simple module interface

In order to achieve this approach, we can carry out the following steps:

1. Structure the module layout with the right files.
2. Create an enum for the different types of orders.
3. Build an order struct.
4. Install the `chrono` crate needed for `datetime` objects.
5. Create an order constructor that utilizes the `chrono` crate.
6. Create dynamic values for the struct.
7. Create a close order interface.
8. Create an open order interface.
9. Use the order interfaces in the main file.

Let's get started:

1. Here, we only allow ourselves to access the stock struct through the order struct. Again, there are other ways to approach this problem, which is a demonstration of how to build interfaces in Rust. In order to achieve this in the code, we have the file structure defined as follows:

```
├── main.rs
└── stocks
    ├── enums
    |   ├── mod.rs
    |   └── order_types.rs
    ├── mod.rs
    └── structs
        ├── mod.rs
        ├── order.rs
        └── stock.rs
```

2. First of all, we can define our enum order types in our enums/order_types.rs file with the following code:

```
pub enum OrderType {
    Short,
    Long
}
```

3. We will use this in our order and interfaces. In order to make this enum type available to the rest of the module, we have to declare it in our enums/mod.rs file with the following code:

```
pub mod order_types;
```

4. Now that we have built our enum type, it is time to put it to work. We can now build our order struct in our stocks/structs/order.rs file with the following code:

```
use chrono::{Local, DateTime};
use super::stock::Stock;
use super::super::enums::order_types::OrderType;

pub struct Order {
```

```
    pub date: DateTime<Local>,
    pub stock: Stock,
    pub number: i32,
    pub order_type: OrderType
}
```

5. Here, we use the `chrono` crate to define when the order was placed; we also have to note what stock the order is for, the number of stocks that we are buying, and the type of order. We have to remember to define our `chrono` dependency on our `Cargo.toml` file with the following code:

```
[dependencies]
serde="1.0.117"
serde_json="1.0.59"
chrono="0.4.19"
```

The reason why we have kept our stock struct separate from the order struct is to allow flexibility. For instance, there are other things that we can do with stock data that is not an order. We may want to build a struct that houses stocks on the user watch list and the user hasn't actually bought anything, but they still want to see the stocks available.

6. There are other use cases for stock data, however. Considering this, we can see that keeping the data and methods around a stock in an individual stock struct helps to not only reduce the amount of code we have to write if we add more features but also standardizes the data around a stock. This also makes it easier for us to maintain the code. If we add or delete a field, or change a method for stock data, we only have to change it in one place as opposed to multiple places. Our constructor for our order struct can be made in the same file with the following code:

```
impl Order {
    pub fn new(stock: Stock, number: i32, \
        order_type: OrderType) -> Order {
        let today: DateTime<Local> = Local::now();
        return Order{date: today, stock, number, \
            order_type}
    }
}
```

Here we create an `Order` struct by accepting `stock`, `number`, and `order_type` arguments and creating a `datetime` struct.

7. Because our order focuses on the logic around pricing the order as it houses the number of stocks brought in an order, in our `impl` block, we can build our current value of the order with the following code:

```
pub fn current_value(&self) -> f32 {
    return self.stock.current_price * self \
        .number as f32
}
```

It has to be noted that we have used `&self` as a parameter instead of just using `self`. This enables us to use the function multiple times. If the parameter was not a reference, then we would move the struct into the function. We would not be able to calculate the value multiple times, and it's going to be useful to do so unless the type is `Copy`.

8. We can also build on this function to calculate the current profit in the `impl` block with the following code:

```
pub fn current_profit(&self) -> f32 {
    let current_price: f32 = self.current_value();
    let initial_price: f32 = self.stock. \
        open_price * self.number as f32;

    match self.order_type {
        OrderType::Long => return current_price -\
            initial_price,
        OrderType::Short => return initial_price -\
            current_price
    }
}
```

Here, we get the current price and the initial price. We then match the order type, as this will change how the profit is calculated. Now our structs are complete, we have to ensure that the structs are available by defining them in the `stocks/structs/mod.rs` file with the following code:

```
pub mod stock;
pub mod order;
```

9. We are now ready to create our interfaces. In order to build our interface in our
    `stocks /mod.rs` file, we initially have to import everything that we need, as
    shown in the following code:

```
pub mod structs;
pub mod enums;

use structs::stock::Stock;
use structs::order::Order;
use enums::order_types::OrderType;
```

10. Now that we have everything to build our interface, we can build our close order
    interface with the following code:

```
pub fn close_order(order: Order) -> f32 {
    println!("order for {} is being closed", \
        &order.stock.name);
    return order.current_profit()
}
```

11. This is a fairly simple interface; we could do more, such as a database or API call,
    but for this demonstration, we merely print that the stock is being sold and return
    the current profit that we have made. With this in mind, we can build our more
    complex interface by opening an order in the same file with the following code:

```
pub fn open_order(number: i32, order_type: OrderType,\
                  stock_name: &str, open_price: f32,\
                  stop_loss: Option<f32>, \
                  take_profit: Option<f32>) -> \
                  Order { \
    println!("order for {} is being made", \
        &stock_name);
    let mut stock: Stock = Stock::new(stock_name, \
        open_price);
    match stop_loss {
        Some(value) => stock = \
            stock.with_stop_loss(value),
        None => {}
    }
```

```
    match take_profit {
        Some(value) => stock = \
            stock.with_take_profit(value),
        None => {}
    }
}
    return Order::new(stock, number, order_type)
}
```

Here, we take in all of the parameters that we need. We have also introduced the `Option<f32>` argument type, which is implemented as an enum type. This allows us to pass in a `None` value. We then create a mutable stock (as the price will vary and we will have to update it), and then check to see whether the `stop_loss` value is provided; if it is, we then add the stop loss to the stock. We then check to see whether the `take_profit` value is provided, updating the stock with this if it is.

12. Now that we have built all our interfaces, all we need to do is to use them in the `main.rs` file. In the main file, we need to import the needed structs and interfaces to utilize them with the following code:

```
mod stocks;

use stocks::{open_order, close_order};
use stocks::structs::order::Order;
use stocks::enums::order_types::OrderType;
```

13. In our main function, we can start putting these interfaces to work by creating a new mutable order with the following code:

```
println!("hello stocks");
let mut new_order: Order = open_order(20, \
    OrderType::Long, "bumper", 56.8, None, None);
```

14. Here, we have set `take_profit` and `stop_loss` to `None`, but we can add them if we need to. To clarify what we have just bought, we can print out the current value and profit with the following code:

```
println!("the current price is: {}",
    &new_order.current_value());
println!("the current profit is: {}",
    &new_order.current_profit());
```

15. We then get some movement in the stock market, which we can simulate by updating the price and printing the value of our investment at each change with the following code:

```
new_order.stock.update_price(43.1);
println!("the current price is: {}", \
    &new_order.current_value());
println!("the current profit is: {}", \
    &new_order.current_profit());

new_order.stock.update_price(82.7);
println!("the current price is: {}", \
        &new_order.current_value());
println!("the current profit is: {}", \
        &new_order.current_profit());
```

16. We now have a profit, and we will sell our stock to close the order and print out the profit with the following code:

```
let profit: f32 = close_order(new_order);
println!("we made {} profit", profit);
```

17. Now, our interfaces, module, and main file are built. Running the Cargo `run` command gives us the following printout:

```
hello stocks
the constructor for the bumper is firing
the current price is: 1136
the current profit is: 0
the current price is: 862
the current profit is: -274
the current price is: 1654
the current profit is: 518
order for bumper is being closed
we made 518 profit
```

As we can see, our module works and it has a clean interface. For this book, our example stops here, as we have shown how we can build modules in Rust with interfaces. However, if you want to go further with building out the application, we can take the approach seen in *Figure 2.6*:

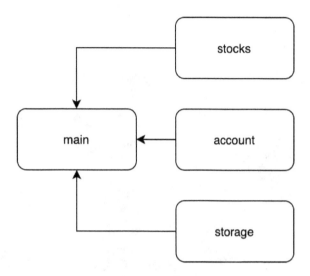

Figure 2.6 – Building out our application

In the account module, we would build data structures around keeping track of the amount the user has through trades. We would then build a storage module that has read and write interfaces for accounts and stocks. The reason why storage is a separate module is that we can keep the interfaces the same, and chop and change the storage logic under the hood.

For instance, we could start with a simple JSON file storage system for development and local usage; however, the application then gets put onto a server, and loads of users start making trades and accessing their accounts. We can switch the file reading and write for a database driver with database model mapping. The system then gets a lot of traffic and the application gets split into a cluster of microservices. One application would still be talking to a database, while another one for frequently requested stocks/accounts could be talking to a Redis cache.

Considering this, keeping the storage separate keeps us flexible. Changing the requirements for the storage is not going to break the build. In fact, a configuration file could enable the switching of different methods, depending on the environment. As long as the interfaces remain the same, refactoring will not be a huge task.

Benefits of documentation when coding

As our module spans multiple files, we are now referencing functions and structs that are in different files. This is where the importance of documentation can be seen. We can revisit our point in the technical requirements of using Visual Studio Code. The code in GitHub is fully documented. If the Rust plugins are installed, merely hovering the mouse over the struct or function will pop up the documentation, allowing us to see what is needed in our interface, as shown in *Figure 2.7*:

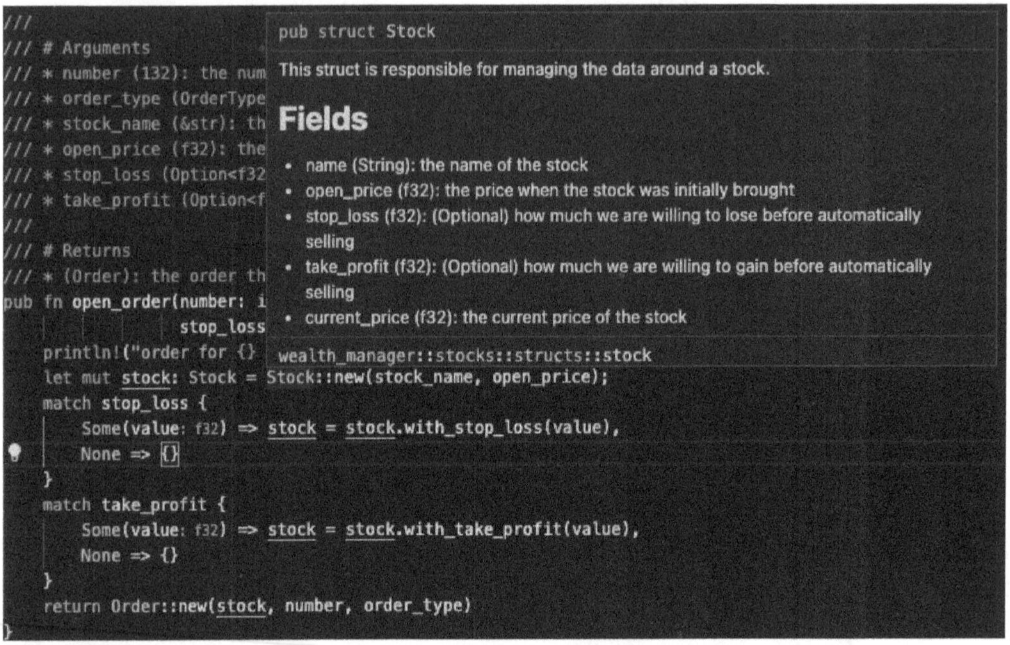

```
///
/// # Arguments
/// * number (132): the num    pub struct Stock
/// * order_type (OrderType     This struct is responsible for managing the data around a stock.
/// * stock_name (&str): th
/// * open_price (f32): the     Fields
/// * stop_loss (Option<f32
/// * take_profit (Option<f        • name (String): the name of the stock
///                               • open_price (f32): the price when the stock was initially brought
/// # Returns                     • stop_loss (f32): (Optional) how much we are willing to lose before automatically
/// * (Order): the order th         selling
pub fn open_order(number: i       • take_profit (f32): (Optional) how much we are willing to gain before automatically
                stop_loss           selling
    println!("order for {}      • current_price (f32): the current price of the stock
    let mut stock: Stock = Stock::new(stock_name, open_price);
    match stop_loss {           wealth_manager::stocks::structs::stock
        Some(value: f32) => stock = stock.with_stop_loss(value),
        None => {}
    }
    match take_profit {
        Some(value: f32) => stock = stock.with_take_profit(value),
        None => {}
    }
    return Order::new(stock, number, order_type)
}
```

Figure 2.7 – Popup documentation in Visual Studio Code

There is a reason why badly structured code that isn't documented is referred to as *tech debt*, and this is because it collects interest over time. Poorly structured code with no documentation is quick to develop, however, and as the size of the application grows, it's going to get harder to change things and understand what is going on. A well-structured module with good Markdown Rust documentation is a great way to keep you and your team's productivity high.

We now have a functioning application that spans multiple pages and is clean and scalable. However, a user cannot dynamically use it, as everything has to be hardcoded. This is not practical. In the next section, we interact with the environment so we can pass arguments into the program.

Interacting with the environment

We are at the stage in which the only thing that is holding us back from building a fully functioning command-line application is interacting with the environment. As stated in the previous section, this is an open-ended subject that spans anything from taking command-line arguments to interacting with servers and databases. As in the previous section, we will cover enough in order to get an understanding of how to structure Rust code that accepts data from the outside and processes it.

In order to explore this, we are going to get our stock application to take in command-line arguments from the user so that we can either buy or sell a stock. We will not over-complicate things by choosing whether to go short or go long, and we will not introduce storage.

However, by the end of this section, we will be equipped to approach building code that scales and accepts data from the outside world. With this, further reading on crates that connect to databases or read/write files will enable us to seamlessly add them to our well-structured code. In terms of databases, we will cover how to mirror the schema of a database and connect to it in *Chapter 10, Injecting Rust into a Python Flask App*.

For our toy example, we will be generating a random number for our sale stock price in order to calculate whether we sell at a profit or loss. We will do this by adding the **rand crate** to our dependencies section in the Cargo.toml file with rand="0.8.3". We can interact with our environment by carrying out the following steps:

1.   Import all the required crates.

2.   Collect the inputs from the environment.

3.   Process our inputs with orders.

Let's get started:

1.   Now that our rand crate has been added, we can add all the extra imports that we need in the main.rs file with the following code:

```
use std::env;
use rand::prelude::*;
use std::str::FromStr;
```

We are using env to get the arguments passed into Cargo. We import everything from the prelude of the rand crate so that we can generate random numbers, and we import FromStr trait so that we can convert strings passed in from the command-line arguments into numbers.

2. In our main function, we initially collect the arguments passed in from the command line with the following code:

```
let args: Vec<String> = env::args().collect();
let action: &String = &args[1];
let name: &String = &args[2];
let amount: i32 = i32::from_str(&args[3]).unwrap();
let price: f32 = f32::from_str(&args[4]).unwrap();
```

We state that we are going to collect the command-line arguments in a vector of strings. We do this because pretty much everything can be represented as a string. We then define all the parameters that we need. We have to note that we start at index 1 instead of 0. This is because index 0 is populated with the run command. We can also see that we are converting the strings into numbers when we need them and directly unwrapping them. This is a little dangerous; we should ideally match the result of the `from_str` function and give better information to the user if we were building a proper production command-line tool.

3. Now that we have everything we need, we create a new order with the data we collected using the following code:

```
let mut new_order: Order = open_order(amount, \
    OrderType::Long, &name.as_str(), price, \
        None, None);
```

We are creating a new order every time even if it is a sell because we do not have storage, and we need to have all the structured data and logic around our stock position. We then match our actions. If we are going to sell our stock, we generate a new price for the stock before selling. Considering this, we can see whether we make a profit or not with the following code:

```
match action.as_str() {
    "buy" => {
        println!("the value of your investment is:\
            {}", new_order.current_value());
    }
    "sell" => {
        let mut rng = rand::thread_rng();

        let new_price_ref: f32 = rng.gen();
        let new_price: f32 = new_price_ref * 100 as \
```

```
            f32;


    new_order.stock.update_price(new_price);
    let sale_profit: f32 = close_order(new_order);

    println!("here is the profit you made: {}", \
        sale_profit);
    }
    _ => {
        panic!("Only 'buy' and 'sell' actions are \
        supported");
    }
}
```

It must be noted that we have a _ at the end of the match expression. This is because the string could theoretically be anything and Rust is a safe language. It will not allow us to compile the code if we did not account for every outcome. The _ is a catch-all pattern. If not all of the match patterns are made, then this is executed. For us, we merely raise an error, stating that only sell and buy are supported.

4. In order to run this program, we perform the following command:

```
cargo run sell monolithai 26 23.4
```

5. Running this will give us the following outcome:

```
order for monolithai is being made
order for monolithai is being closed
here is the profit you made: 1825.456
```

The profit you make will be different, as the number generated will be random.

Here we have it – our application is interactive and scalable. If you want to build more comprehensive command-line interfaces with help menus, it is recommended that you read and utilize the clap crate.

Summary

In this chapter, we went through the basics of Cargo. With Cargo, we managed to build basic applications, document them, compile them, and run them. Looking at how clean and easy this implementation was, it is clear to see why Rust is one of the most favored languages. Managing all the functionality, documentation, and dependencies in one file with a few lines of code speeds up the whole process. Combining this with a strict, helpful compiler makes Rust a no-brainer when it comes to managing complex projects. We managed our complexity by wrapping our module in easy-to-use interfaces and interacting with the user's inputs through the command line.

Right now, as you stand, you can start building Rust code to solve a range of problems. If you want to build an application that interacts as a Rust web server with a frontend and database, I recommend that you read my other book on web development in Rust, *Rust Web Programming*, and start at *Chapter 3*, as you have now covered enough Rust fundamentals to start building Rust servers.

In the next chapter of this book, we will cover how to exploit Rust's concurrency.

Questions

1. As we continue to code, how do we document it?

2. Why is it important to keep modules isolated to a single concept?

3. How do we enable our modules to keep the advantages of isolated modules?

4. How do we manage dependencies in our application?

5. How do we ensure that all outcomes in a match expression are accounted for when there is theoretically an infinite number of outcomes, such as matching different strings?

6. Let's say that we have a struct called `SomeStruct` in a `some_file/some_struct.rs` file. How do we make this available outside of the directory that it is in?

7. Let's say that we have changed our mind about our `SomeStruct` struct in question 6 and we want it only available in the `some_file/` directory. How would we do this?

8. How can we access our `SomeStruct` struct in the `some_file/another_struct.rs` file?

Answers

1. Our docstrings can support Markdown while we are building our structs and functions. Because it's Markdown, we can document ways in which we can implement the struct or function. If we are using Visual Studio Code, this also helps our productivity, as merely hovering the mouse over the function or struct throws up the documentation.

2. Keeping our modules constrained to a single concept increases the flexibility of the application, enabling us to chop and change modules as and when they are needed.

3. In order to keep our modules isolated, we need to keep the interfaces of the module the same; this means that we can change logic inside the module without having to alter anything in the rest of the application. If we delete the module, we only have to look for implementations of the interface throughout the application as opposed to the implementation of all functions and structs in the module.

4. We manage our dependencies in the `Cargo.toml` file. Just running Cargo will install the requirements we have when it is compiling before running.

5. We can account for everything by catching anything that hasn't satisfied all our matches. This is done by implementing a _ pattern at the end of our match expression, executing the code attached to that.

6. We make it publicly available by writing `pub mod some_struct;` in the `some_file/mod.rs` file.

7. We make it available only in the `some_file/` directory by writing `mod some_struct;` in the `some_file/mod.rs` file.

8. We can access the `SomeStruct` by typing `use super::some_struct::SomeStruct;` in the `some_file/another_struct.rs` file.

Further reading

- *Rust Web Programming, Maxwell Flitton, Packt Publishing* (2021)

- *Mastering Rust, Rahul Sharma and Vesa Kaihlavirta, Packt Publishing* (2019)

- *The Rust Programming Language,* Rust Foundation: `https://doc.rust-lang.org/stable/book/` (2018)

- *The Clap documentation,* Clap Docs: `https://docs.rs/clap/2.33.3/clap/` (2021)

- *The standard file documentation,* Rust Foundation: `https://doc.rust-lang.org/std/fs/struct.File.html` (2021)

- *The chrono DateTime documentation,* Rust Foundation: `,https://docs.rs/chrono/0.4.19/chrono/struct.DateTime.html` (2021)

3
Understanding Concurrency

Speeding up our code with Rust is useful. However, understanding concurrency and utilizing threads and processes can take our ability to speed up our code to the next level. In this chapter, we will go through what processes and threads are. We then go through the practical steps of spinning up threads and processes in Python and Rust. However, while this can be exciting, we also must acknowledge that reaching for threads and processes without thinking about our approach can end up tripping us up. To avoid this, we also explore algorithm complexity and how this affects our computation time.

In this chapter, we will cover the following topics:

- Introducing concurrency
- Basic asynchronous programming with threads
- Running multiple processes
- Customizing threads and processes safely

Technical requirements

The code for this chapter can be accessed via the following GitHub link:

`https://github.com/PacktPublishing/Speed-up-your-Python-with-Rust/tree/main/chapter_three`

Introducing concurrency

As we explored in the introduction of *Chapter 1, An Introduction to Rust from a Python Perspective*, Moore's law is now failing, and therefore we have to consider other ways in which we can speed up our processing. This is where concurrency comes in. Concurrency is essentially running multiple computations at the same time. Concurrency is everywhere, and to give the concept full justice, we would have to write a whole book on it.

However, for the scope of this book, understanding the basics of concurrency (and when to use it) can add an extra tool to our belt that enables us to speed up computations. Furthermore, threads and processes are how we can break up our program into computations that run at the same time. To start our concurrency tour, we will cover threads.

Threads

Threads are the smallest unit of computation that we can process and manage independently. Threads are used to break a program into computational parts that can be run at the same time. It also has to be noted that threads can be run out of sequence. This brings forward an important distinction between concurrency and parallelism. **Concurrency** is the task of running and managing multiple computations at the same time, while **parallelism** is the task of running multiple computations at the same time. Concurrency has a non-deterministic control flow, while parallelism has a deterministic control flow. Threads share resources such as memory and processing power; however, they also block each other. For instance, if we spin off a thread that requires constant processing power, we will merely block the other thread, as seen in the following diagram:

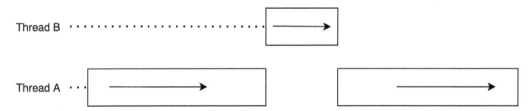

Figure 3.1 – Two threads over time

Here, we can see that **Thread A** stops running when **Thread B** is running. This is demonstrated in Pan Wu's 2020 article on understanding multithreading through simulations where different types of tasks were timed. The results in the article are summed up in the following chart:

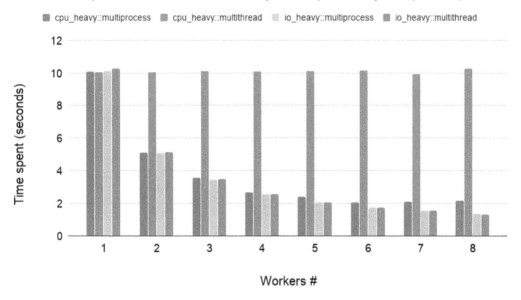

Figure 3.2 – Times of different tasks [Source: Pan Wu, https://towardsdatascience.com/understanding-python-multithreading-and-multiprocessing-via-simulation-3f600dbbfe31]

Here, we can see that the times decrease as the number of workers decreases, *apart from* the **central processing unit** (**CPU**)-heavy multithreaded tasks. This is because, as demonstrated in *Figure 3.1*, the CPU-intensive threads are blocking, so only one worker can process at a time. It does not matter how many more workers you add. It must be noted that this is because of Python's **global interpreter lock** (**GIL**), which is covered in *Chapter 6*, *Working with Python Objects in Rust*. In other contexts, such as Rust, they can be executed on different CPU cores and generally will not block each other.

We can also see in *Figure 3.2* that the **input/output** (**I/O**)-heavy tasks do reduce in time taken when the workers increase. This is because there is idle time in I/O-heavy tasks. This is where we can really utilize threads. Let's say our task is making a call to a server. There is some idle time when waiting for a response, therefore utilizing threads to make multiple calls to servers will speed up the time. We also must note that processes work for CPU- and I/O-heavy tasks. Because of this, it is beneficial for us to explore what processes are.

Processes

Processes are more expensive to produce compared to threads. In fact, a process can host multiple threads. This is usually depicted in the following classic multithreading diagram, as seen everywhere (including the multiprocessing *Wikimedia* page):

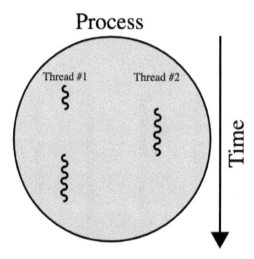

Figure 3.3 – Relationship between threads and processes [Source: Cburnett (2007) (https://commons.wikimedia.org/wiki/File:Multithreaded_process.svg), CC BY-SA 3.0]

This is a classic diagram because it encapsulates the relationship between processes and threads so well. Here, we can see that threads are a subset of a process. We can also see why threads share memory, and as a result, we must note that processes are typically independent and do not share memory. We also must note that context switches are more expensive when using processes. A context switch refers to when the state of a process (or thread) is stored so that it can be restored and resumed at a later state. An example of this would be waiting for an **application programming interface** (**API**) response. The state can be saved, and another process/thread can run while we wait for the API response.

Now that we understand the basic concepts behind threads and processes, we need to learn how to practically use threads in our programs.

Basic asynchronous programming with threads

To utilize threading, we need to be able to start threads, allow them to run, and then join them. We can see the stages of practically managing our threads in the following diagram:

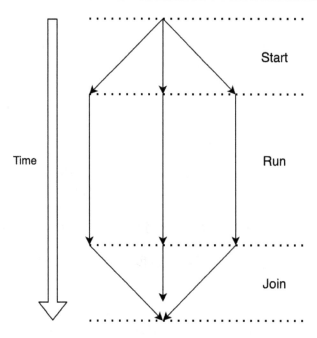

Figure 3.4 – Stages of threads

We start the threads, we then let them run, and once they have run, we join them. If we did not join them, the program would continue to run before the threads had finished. In Python, we create a thread by inheriting the Thread object, as follows:

```python
from threading import Thread
from time import sleep
from typing import Optional


class ExampleThread(Thread):

    def __init__(self, seconds: int, name: str) -> None:
        super().__init__()
        self.seconds: int = seconds
        self.name: str = name
        self._return: Optional[int] = None

    def run(self) -> None:
        print(f"thread {self.name} is running")
```

```
        sleep(self.seconds)
        print(f"thread {self.name} has finished")
        self._return = self.seconds

    def join(self) -> int:
        Thread.join(self)
        return self._return
```

Here, we can see that we have overwritten the run function in the Thread class. This function runs when the thread is running. We then overwrite the join method. However, we must note that in the join function, there is extra functionality going on under the hood; therefore, we must call the Thread class's join method, and then return whatever we want at the end. We do not have to return anything if we do not want to. If this is the case, then there is no point overwriting the join function. We can then implement the threads by running the following code:

```
one: ExampleThread = ExampleThread(seconds=5, name="one")
two: ExampleThread = ExampleThread(seconds=5, name="two")
three: ExampleThread = ExampleThread(seconds=5,
    name="three")
```

We then have to time the process of starting, running, and joining the outcomes, like so:

```
import time

start = time.time()
one.start()
two.start()
three.start()
print("we have started all of our threads")
one_result = one.join()
two_result = two.join()
three_result = three.join()
finish = time.time()
print(f"{finish - start} has elapsed")
print(one_result)
print(two_result)
print(three_result)
```

When we run this code, we get the following console printout:

```
thread one is running
thread two is running
thread three is running
we have started all of our threads
thread one has finished
thread three has finished
thread two has finished
5.005641937255859 has elapsed
5
5
5
```

Straight away, we can see that it took just over 5 seconds to execute the whole process. If we were running our program sequentially, it would take 15 seconds. This shows that our threads are working!

It also must be noted that thread `three` finished before thread `two`, even though thread `two` started before. Don't worry if you get a finishing sequence of `one`, `two`, `three`; this is because threads finish in an indeterminate order. Even though the scheduling is deterministic, there are thousands of events and processes running under the hood of the CPU when the program is running. As a result, the exact time slices that each thread gets are never the same. These tiny changes add up over time, and as a result, we cannot guarantee that the threads will finish in a determinate order if the executions are close and the durations are roughly the same.

Now we have the basics of Python threads, we can move on to spinning off threads in Rust. However, before we start doing this, we must understand the concept of **closures**, which are essentially a way to store functions anonymously along with the environment it is part of. Considering this, we can define functions within the scope of the `main` function or inside other scopes including other functions. A simple example of a building closure is to print an input, like this:

```rust
fn main() {
    let example_closure: fn(&str) = |string_input: &str| {
        println!("{}", string_input);
    };
    example_closure("this is a closure");
}
```

With this approach, we can exploit scopes. It also must be noted that as closures are scope-sensitive, we can also utilize the existing variables around a closure. To demonstrate this, we can create a closure that calculates the amount of interest we have to pay on a loan due to the external base rate. We will also define it in an inner scope, as seen here:

```
fn main() {
        let base_rate: f32 = 0.03;
        let calculate_interest = |loan_amount: &f32| {
            return loan_amount * &base_rate
        };
        println!("the total interest to be paid is: {}",
            calculate_interest(&32567.6));
}
```

Running this code would give us the following printout in the console:

```
the total interest to be paid is: 977.02795
```

Here, we can see that closures can return values, but we have not defined the type for the closure. This is the case even though it is returning a float. In fact, if we set `calculate_interest` to `f32`, the compiler would complain, stating that the types were mismatched. This is because the closure is a unique anonymous type that cannot be written out. A closure is a struct generated by the compiler that houses captured variables. If we try to call the closure outside the inner scope, our application will fail to compile as the closure cannot be accessed outside the scope.

Now that we have covered Rust closures, we can replicate the Python threading example that we covered earlier in the section. Initially, we must import the standard module crates that are required by running the following code:

```
use std::{thread, time};
use std::thread::JoinHandle;
```

We are using `thread` to spawn off threads, `time` to keep track of how long our processes take, and the `JoinHandle` struct to join the threads. With these imports, we can build our own thread by running the following code:

```
fn simple_thread(seconds: i8, name: &str) -> i8 {
    println!("thread {} is running", name);
    let total_seconds = time::Duration::new(seconds as \
        u64, 0);
```

```
    thread::sleep(total_seconds);
    println!("thread {} has finished", name);
    return seconds
}
```

Here, we can see that we create a `Duration` struct denoted as `total_seconds`. We then use the thread and `total_seconds` to put the function to sleep, returning the number of seconds when the whole process is finished. Right now, this is just a function, and running it by itself will not spin off different threads. Inside our `main` function, we start our timer and spawn off our three threads by running the following code:

```
let now = time::Instant::now();

let thread_one: JoinHandle<i8> = thread::spawn(|| {
    simple_thread(5, "one")});
let thread_two: JoinHandle<i8> = thread::spawn(|| {
    simple_thread(5, "two")});
let thread_three: JoinHandle<i8> = thread::spawn(|| {
    simple_thread(5, "three")});
```

Here, we spawn threads and pass our function in with the right parameters in the closure. Nothing is stopping us from putting any code in the closure. The final line in the closure would be what is returned to the `JoinHandle` struct to unwrap. Once this is done, we join all the threads to hold the program until all the threads have finished before moving on with this code:

```
let result_one = thread_one.join();
let result_two = thread_two.join();
let result_three = thread_three.join();
```

The `join` function returns a result with the `Result<i8, Box<dyn Any + Send>>` type.

There are some new concepts here, but we can break them down, as follows:

- We remember that a `Result` struct in Rust either returns an `Ok` or an `Err` response. If the thread runs without any problems, then we will return the `i8` value that we are expecting. If not, then we have this rather ugly `Result<i8, Box<dyn Any + Send>>` output as the error.

- The first thing we must address here is the Box struct. This is one of the most basic forms of a pointer and allows us to store data on the heap rather than the stack. What remains on the stack is the pointer to the data in the heap. We are using this because we do not know how big the data is when coming out of the thread.

- The next expression that we must explain is dyn. This keyword is used to indicate that the type is a trait object. For instance, we might want to store a range of Box structs in an array. These Box structs might point to different structs. However, we can still ensure that they can be grouped together if they have a certain trait in common. For instance, if all the structs had to have TraitA implemented, we would denote this with Box<dyn TraitA>.

- The Any keyword is a trait for dynamic typing. This means that the data type can be anything. The Any trait is combined with Send by using the Any + Send expression. This means that both traits must be implemented.

- The Send trait is for types that can be transferred across thread boundaries. Send is implemented automatically by the compiler if it is deemed appropriate. With all this, we can confidently state that the join of a thread in Rust returns a result that can either be the integer that we desire or a pointer to anything else that can be transferred across threads.

To process the results of the thread, we could just directly unwrap them. However, this would not be very useful when the demands of our multithreaded programs increase. We must be able to handle what potentially comes out of a thread, and to do this, we are going to have to downcast the outcome. Downcasting is Rust's method of converting a trait into a concrete type. In this context, we will be converting PyO3 structs that denote Python types into concrete Rust data types such as strings or integers. To demonstrate this, let's build a function that handles the outcome of our thread, as follows:

1. First, we are going to have to import everything we need, as seen in the following code snippet:

```
use std::any::Any;
use std::marker::Send;
```

2. With these imports, we can create a function that unpacks the result and prints it using this code:

```
fn process_thread(thread_result: Result<i8, Box<dyn \
    Any + Send>>, name: &str) {
    match thread_result {
        Ok(result) => {
```

```
            println!("the result for {} is {}", \
                result, name);
        }
        Err(result) => {
            if let Some(string) = result.downcast \
            _ref::<String>() {
                println!("the error for {} is: {}", \
                    name, string);
            } else {
                println!("there error for {} does \
                    not have a message", name);
            }
        }
    }
}
```

3. Here, we merely print out the result if it is a success. However, if it is an error, as pointed out earlier, we do not know what data type the error is. However, we would still like to handle this. This is where we downcast. Downcasting returns an option, which is why we have the `if let Some(string) = result.downcast_ref::<String>()` condition. If the downcast is successful, we can move the string into the scope and print out the error string. If it is not successful, we can move on and state that although there was an error, an error string was not provided. We can use multiple conditional statements to account for a range of data types if we want. We can write a lot of Rust code without having to rely on downcasting, as Rust has a strict typing section. However, when interfacing with Python this can be useful, as we know that Python objects are dynamic and could essentially be anything.

4. Now that we can process our threads when they have finished, we can stop the clock and process the outcomes by running the following code:

```
println!("time elapsed {:?}", now.elapsed());
process_thread(result_one, "one");
process_thread(result_two, "two");
process_thread(result_three, "three");
```

5. This gives us the following printout:

```
thread one is running
thread three is running
thread two is running
thread one has finished
thread three has finished
thread two has finished
time elapsed 5.00525725s
the result for 5 is one
the result for 5 is two
the result for 5 is three
```

And here we have it: we can run and process threads in Python and Rust. However, remember that if we try to run CPU-intensive tasks with the code that we have written, we will not get the speed up. However, it must be noted that in the Rust context of the code, there could be a speedup depending on the environment. For instance, if multiple CPU cores are available, the **operating system (OS)** scheduler can put those threads onto those cores to be executed in parallel. To write code that will speed up our code in this context, we will have to learn how to practically spin up multiple processes, which we cover in the next section.

Running multiple processes

Technically with Python, we can simply switch the inheritance of our thread from Thread to Process by running the following code:

```python
from multiprocessing import Process
from typing import Optional


class ExampleProcess(Process):

    def __init__(self, seconds: int, name: str) -> None:
        super().__init__()
        self.seconds: int = seconds
```

```
        self.name: str = name
        self._return: Optional[int] = None

    def run(self) -> None:
        # do something demanding of the CPU
        pass

    def join(self) -> int:
        Process.join(self)
        return self._return
```

However, there are some compilations. If we refer to *Figure 3.3*, we can see that processes have their own memory. This is where things can get complicated.

For instance, there is nothing wrong with the process defined previously if the process is not returning anything directly but writing to a database or file. On the other hand, the join function will not return anything directly and will just have None instead. This is because Process is not sharing the same memory space as the main process. We also must remember that spinning off processes is more expensive, so we must be more careful with this.

Since we are getting more complex with the memory and the resources are getting more expensive, it makes sense to rein it in and keep it simple. This is where we utilize a **pool**. A pool is where we have several workers processing inputs simultaneously and then packaging them as an array, as seen here:

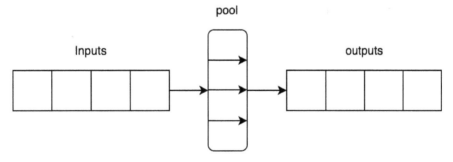

Figure 3.5 – Pool of processes

The advantage here is that we keep the expensive multiprocessing context to a small part of the program. We can also easily control the number of workers that we are willing to support. For Python, this means that we keep the interaction as lightweight as possible. As seen in the next diagram, we package an individual isolated function in a tuple with an array of inputs. This tuple gets processed in the pool by a worker, and then the result of the outcome is returned from the pool:

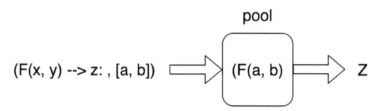

Figure 3.6 – Pool data flow

To demonstrate multiprocessing via a pool, we can utilize the **Fibonacci sequence**. This is where the next number of the sequence is the sum of the previous number in the sequence and the number before that, as illustrated here:

$$F_n = F_{n-1} + F_{n-2}$$

To calculate a number in the sequence, we will have to use **recursion**. There is a closed form of the Fibonacci sequence; however, this will not let us explore multiprocessing as the closed sequence by its very nature doesn't scale in computation as n increases. To calculate a Fibonacci number in Python, we can write an isolated function, as seen in the following code snippet:

```python
def recur_fibo(n: int) -> int:
    if n <= 1:
        return n
    else:
        return (recur_fibo(n-1) + recur_fibo(n-2))
```

This function keeps going back until it hits the bottom of the tree at either 1 or 0. This function is terrible at scaling. To demonstrate this, let's look at the recursion tree shown here:

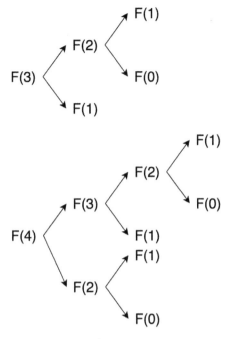

Figure 3.7 – Fibonacci recursion tree

We can see that these are not perfect trees, and if you go online and search for *big O notation of the Fibonacci sequence*, there are debates, and some equations will equate the scaling factor to the golden ratio. While this is interesting, it is outside the scope of this book as we are focusing on computational complexity. As a result, we will simplify the math and treat this as a perfectly symmetrical tree. Recursion trees scale at the rate of 2^n, where n is the depth of the tree. Referring to *Figure 3.7*, we can see that if we treat the tree as perfectly symmetrical, a n value of 3 has a depth of 3, and a n value of 4 has a depth of 4. As n increases, the computation increases exponentially.

We have taken a slight detour of complexity to highlight the importance of taking this into account before reaching for multiprocessing. The reason why you bought this book as opposed to searching online for multiprocessing code snippets to copy and paste into your code is that you want to be guided through these concepts with pointers for further reading and an appreciation of their context. In the case of this sequence, reaching for a closed form or caching answers would reduce the computation time greatly. If we have an ordered list of numbers, getting the highest number in the list and then creating a full sequence up to the highest number would be a lot quicker than repeatedly calculating the sequence again and again for each number we want to calculate. Avoiding recursion altogether is a better option than reaching for multiprocessing.

To implement and test our multiprocessing pool, we first need to time how long it would take to calculate a range of numbers sequentially. This can be done like so:

```python
import time

start = time.time()
recur_fibo(n=8)
recur_fibo(n=12)
recur_fibo(n=12)
recur_fibo(n=20)
recur_fibo(n=20)
recur_fibo(n=20)
recur_fibo(n=20)
recur_fibo(n=28)
recur_fibo(n=28)
recur_fibo(n=28)
recur_fibo(n=28)
recur_fibo(n=36)
finish = time.time()
print(f"{finish - start} has elapsed")
```

We have introduced a pretty long list; however, this is essential to see the difference. If we just had two Fibonacci numbers to compute, then the cost of spinning up processes could eclipse the gain in multiprocessing.

Our multiple processing pool can be implemented as follows:

```python
if __name__ == '__main__':
    from multiprocessing import Pool

    start = time.time()
    with Pool(4) as p:
        print(p.starmap(recur_fibo, [(8,), (12,), (12,), \
            (20,), (20,), (20,), (20,), (28,), (28,), (28,), \
            (28,),(36,)]))
    finish = time.time()
    print(f"{finish - start} has elapsed")
```

Please note that we have nested this code under if __name__ == "__main__":. This is because the whole script gets run again when spinning up another process, which can result in infinite loops. If the code is nested under if __name__ == "__main__": then it will not run again as there is only one main process. It also must be noted that we defined a pool of four workers. This can be changed to whatever we feel fit but there are diminishing returns when increasing this, as we will explore later. The tuples in the list are the arguments for each computation. Running the whole script gives us the following output:

```
3.2531330585479736 has elapsed
[21, 144, 144, 6765, 6765, 6765, 6765, 317811,
317811, 317811, 317811, 14930352]
3.100019931793213 has elapsed
```

We can see that the speed is not a quarter of the sequential calculations. However, the multiprocessing pool is slightly faster. If you run this multiple times, you will get some variance in the difference in times. However, the multiprocessing approach will always be faster. Now that we have run a multiprocessing tool in Python, we can implement our Fibonacci multithreading in the different context of a multiprocessing pool in Rust. Here's how we'll do this:

1. In our new Cargo project, we can code the following function in our main.rs file:

```rust
pub fn fibonacci_recursive(n: i32) -> u64 {
    if n < 0 {
        panic!("{} is negative!", n);
    }
    match n {
        0       => panic!(
        "zero is not a right argument to
        fibonacci_reccursive()!"),
        1 | 2 => 1,
        _       => fibonacci_reccursive(n - 1) +
                fibonacci_reccursive(n - 2)
    }
}
```

We can see that our Rust function is not more complex than our Python version. The extra lines of code are just to account for unexpected inputs.

2. To run this and time it, we must import the time crate at the top of the `main.rs` file by running the following code:

```
use std::time;
```

3. Then, we must compute the exact same Fibonacci numbers as we did in the Python implementation, as follows:

```
fn main() {
    let now = time::Instant::now();
    fibonacci_reccursive(8);
    fibonacci_reccursive(12);
    fibonacci_reccursive(12);
    fibonacci_reccursive(20);
    fibonacci_reccursive(20);
    fibonacci_reccursive(20);
    fibonacci_reccursive(20);
    fibonacci_reccursive(28);
    fibonacci_reccursive(28);
    fibonacci_reccursive(28);
    fibonacci_reccursive(28);
    fibonacci_reccursive(36);
    println!("time elapsed {:?}", now.elapsed());
}
```

4. To run this, we are going to use the following command:

```
cargo run -release
```

5. We are going to use the release version as that is what we will be using in production. Running it gives us the following output:

```
time elapsed 40.754875ms
```

Running this several times will give us an average roundabout of 40 milliseconds. Considering that our multiprocessing Python code ran at roughly 3.1 seconds, our Rust single-threaded implementation runs 77 times faster than our Python multiprocessing code. Just let that sink in! The code was not more complex, and it is memory-safe. Therefore, fusing Rust with Python is such a quick win! Combining the aggressive typing with the compiler forces us to account for every input and outcome, and we are on the way to turbocharging our Python systems with safer, faster code.

Now, we are going to see what happens to the speed when we run our numbers through a multithreading tool. Here's how we'll go about this:

1. To do this, we are going to use the `rayon` crate. We define this dependency in our `Cargo.toml` file by running the following code:

    ```
    [dependencies]
    rayon="1.5.0"
    ```

2. Once this is done, we import it into our `main.rs` file, as follows:

    ```
    use rayon::prelude::*;
    ```

3. We can then run our multithreading pool in our `main` function below our sequential calculations by running the following code:

    ```
    rayon::ThreadPoolBuilder::new().num_threads(4) \
        .build_global().unwrap();

    let now = time::Instant::now();
    let numbers: Vec<i32> = vec![8, 12, 12, 20, 20, 20, \
        20, 28, 28, 28, 28, 36];
    let outcomes: Vec<u64> = numbers.into_par_iter() \
        .map(|n| fibonacci_reccursive(n)).collect();
    println!("{:?}", outcomes);
    println!("time elapsed {:?}", now.elapsed());
    ```

4. Here, we define the number of threads that our pool builder has. We then execute the `into_par_iter` function on the vector. This is achieved by implementing the `IntoParallelIterator` trait onto the vector, which is done when the `rayon` crate is imported. If it were not imported, then the compiler would complain, stating that a vector does not have the `into_par_iter` function associated with it.

5. We then map our Fibonacci function over the integers in the vector utilizing a closure and collect them. The calculated Fibonacci numbers are associated with the `outcomes` variable.

6. We then print them and print the time elapsed. Running this via a release gives us the following printout in the console:

    ```
    time elapsed 38.993791ms
    [21, 144, 144, 6765, 6765, 6765, 6765, 317811,
    ```

```
317811, 317811, 317811, 14930352]
time elapsed 31.493291ms
```

Running this several times will give you roughly the times stated in the preceding console printout. Calculating this gives us a 20% increase in speed. Considering that the Python multiprocessing only gave us a 5% increase, we can deduce that Rust is also more efficient at multithreading when the right context is applied.

We can go a little further to really see the advantages of these pools. Remember that our sequence increases exponentially. In our Rust program, we can add three computations for n being 46 to our sequential calculations and pooled calculations, and we get the following output:

```
time elapsed 12.5856675s
[21, 144, 144, 6765, 6765, 6765, 6765, 317811, 317811,
317811, 317811, 14930352, 1836311903, 1836311903,
1836311903]
time elapsed 4.0485755s
```

First, we must acknowledge that the time went from milliseconds to double-digit seconds. Exponential scaling algorithms are painful, and just adding 10 to your calculation pushes it up greatly. We can also see that our savings have increased. Our pooled calculations are now 3.11 times faster as opposed to 1.2 times faster in the previous test!

7. If we add three extra computations for n being 46 for our Python implementation, we get the following console printout:

```
1105.5351197719574 has elapsed
[21, 144, 144, 6765, 6765, 6765, 6765, 317811, 317811,
317811, 317811, 14930352, 1836311903, 1836311903,
1836311903]
387.0687129497528 has elapsed
```

Here, we can see that our Python pooled processing is 2.85 times faster than our Python sequential processing. We also must note here that our Rust sequential processing is roughly 95 times faster than our Python sequential processing, and our Rust pool multithreading is roughly 96 times faster than our Python pool processing. As the number of points that need processing increases, so will the difference. This highlights even more motivation for plugging Rust into Python.

It must be noted that we got our speed increase in our Rust program through multithreading as opposed to multiprocessing. Multiprocessing in Rust is not as straightforward as in Python—this is mainly down to Rust being a newer language. For instance, there is a crate called mitosis that will enable us to run functions in a separate process; however, this crate only has four contributors, and the last contribution at the time of writing this book was 13 months ago. Considering this, we should approach multiprocessing in Rust without any third-party crates. To achieve this, we need to code a Fibonacci calculation program and a multiprocessing program that will call this in different processes, as seen in the following diagram:

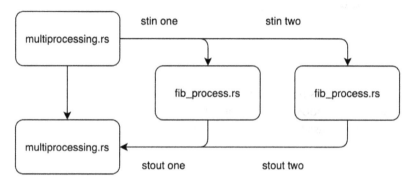

Figure 3.8 – Multiprocessing in Rust

We are going to pass our data into these processes and parse the outputs handling them in our multiprocessing.rs file. To carry this out in the simplest way, we code both files in the same directory. First, we build our fib_process.rs file. We must import what we are going to do by running the following code:

```
use std::env;
use std::vec::Vec;
```

We want our processes to accept a list of integers to calculate, so we define Fibonacci number and numbers functions, as follows:

```
pub fn fibonacci_number(n: i32) -> u64 {
    if n < 0 {
        panic!("{} is negative!", n);
    }
    match n {
        0       => panic!("zero is not a right argument \
                        to fibonacci_number!"),
        1 | 2 => 1,
```

```
            _       => fibonacci_number(n - 1) +
                       fibonacci_number(n - 2)
        }
    }
}
pub fn fibonacci_numbers(numbers: Vec<i32>) -> Vec<u64> {
    let mut vec: Vec<u64> = Vec::new();

    for n in numbers.iter() {
        vec.push(fibonacci_number(*n));
    }

    return vec
}
```

We have seen these functions before as they have become the standard way to calculate Fibonacci numbers in this book. We now must take a list of integers from arguments, parse them into integers, pass them into our calculation function, and return the results, as follows:

```
fn main() {
    let mut inputs: Vec<i32> = Vec::new();
    let args: Vec<String> = env::args().collect();
    for i in args {
        match i.parse::<i32>() {
            Ok(result) => inputs.push(result),
            Err(_) => (),
        }
    }
    let results = fibonacci_numbers(inputs);
    for i in results {
        println!("{}", i);
    }
}
```

Here, we can see that we collect the input from the environment. Once the input integers have been parsed into i32 integers and used to calculate the Fibonacci numbers, we merely print them out. Printing out to the console generally acts as stdout. Our process file is fully coded, so we can compile it with the following command:

```
rustc fib_process.rs
```

This creates a binary of our file. Now that this is done, we can move on to our multiprocessing.rs file that will spawn multiple processes. We import what we need by running the following code:

```
use std::process::{Command, Stdio, Child};
use std::io::{BufReader, BufRead};
```

The Command struct is going to be used to spawn off a new process, the Stdio struct is going to be used to define the piping of data back from the process, and the Child struct is returned when the process is spawned. We will use them to access the output data and get the process to wait to finish. The BufReader struct is used to read the data from the child process. Now that we have imported everything we need, we can define a function that accepts an array of integers as strings and spins off the process, returning the Child struct, as follows:

```
fn spawn_process(inputs: &[&str]) -> Child {
    return Command::new("./fib_process").args(inputs)
    .stdout(Stdio::piped())
    .spawn().expect("failed to execute process")
}
```

Here, we can see that we just must call our binary and pass in our array of strings in the args function. We then define the stdout and spawn the process, returning the Child struct. Now that this is done, we can fire off three processes in our main function and wait for them to complete by running the following code:

```
fn main() {
    let mut one = spawn_process(&["5", "6", "7", "8"]);
    let mut two = spawn_process(&["9", "10", "11", "12"]);
    let mut three = spawn_process(&["13", "14", "15", \
        "16"]);

    one.wait();
    two.wait();
```

```
        three.wait();
}
```

We can now start extracting the data from these processes inside our `main` function by running the following code:

```
let one_stdout = one.stdout.as_mut().expect(
    "unable to open stdout of child");
let two_stdout = two.stdout.as_mut().expect(
    "unable to open stdout of child");
let three_stdout = three.stdout.as_mut().expect
("unable to open stdout of child");


let one_data = BufReader::new(one_stdout);
let two_data = BufReader::new(two_stdout);
let three_data = BufReader::new(three_stdout);
```

Here, we can see that we have accessed the data using the `stdout` field, and then processed it using the `BufReader` struct. We can then loop through our extracted data, appending it to an empty vector and then printing it out by running the following code:

```
let mut results = Vec::new();
for i in three_data.lines() {
    results.push(i.unwrap().parse::<i32>().unwrap());
}
for i in one_data.lines() {
    results.push(i.unwrap().parse::<i32>().unwrap());
}
for i in two_data.lines() {
    results.push(i.unwrap().parse::<i32>().unwrap());
}
println!("{:?}", results);
```

This code is a little repetitive, but it illustrates how to spawn and manage multiple processes in Rust. We then compile the file with the following command:

```
rustc fib_multiprocessing.rs
```

We can then run our multiprocessing code with the following command:

```
./multiprocessing
```

We then get the output, as follows:

```
[233, 377, 610, 987, 5, 8, 13, 21, 34, 55, 89, 144] we have
    it, our multiprocessing code in Rust works.
```

We have now covered all we need to know about running processes and threads to speed up our computations. However, we need to be mindful and investigate how to customize our threads and processes safely to avoid pitfalls.

Customizing threads and processes safely

In this section, we will cover some of the pitfalls that we have to avoid when being creative with threads and processes. We will not cover the concepts in depth, as advanced multiprocessing and concurrency is a big topic and there are books completely dedicated to this. However, it is important to understand what to look out for and which topics to read if you want to increase your knowledge of multiprocessing/threading.

Looking back at our Fibonacci sequences, it might be tempting to spin off extra threads inside our thread to speed up the individual computations in the thread pool. However, to truly understand if this is a good idea, we need to understand **Amdahl's law**.

Amdahl's law

Amdahl's law lets us describe the trade-off on adding more threads. If we spin off threads inside the threads, we will have exponential growth of threads. You may be forgiven for thinking this to be a good idea; however, Amdahl's law states that there are diminishing returns when increasing the cores. Have a look at the following formula:

$$Speed_{latency}(s) = \frac{1}{(1-p) + \frac{p}{s}}$$

Here, the following applies:

- *Speed*: This is the theoretical speedup of the execution of the whole task.
- *s*: This is the speedup of the part of the task that benefits from improved system resources.
- *p*: This is the proportion of execution time that the part benefiting from improved resources originally occupied.

In general, increasing the cores does have an impact; however, the diminishing returns can be seen in the following screenshot:

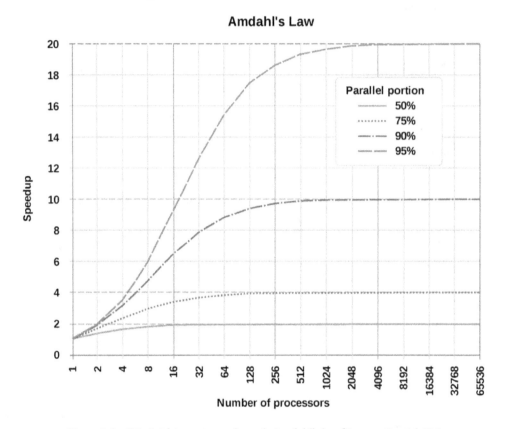

Figure 3.9 – Diminishing returns through Amdahl's law [Source: Daniels220 (https://commons.wikimedia.org/w/index.php?curid=6678551), CC BY-SA 3.0]

Considering this, we might want to investigate using a broker to manage our multiprocessing. However, this can lead to using **clogging** up the broker, resulting in **deadlock**. To understand the gravity of this situation, we will explore deadlocks in the next section.

Deadlocks

Deadlocks can arise when it comes to bigger applications, where it is common to manage the multiprocessing through a task broker. This is usually managed via a database or caching mechanism such as Redis. This consists of a queue where the tasks are added, as illustrated here:

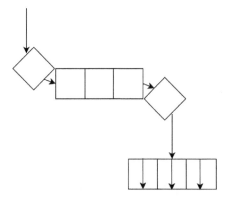

Figure 3.10 – The flow of tasks when multiprocessing with a broker or queue

Here, we can see that new tasks can be added to the queue. As time goes on, the oldest tasks get taken off the queue and passed into the pool. Throughout the application, our code can send functions and parameters to the queue anywhere in the application.

In Python, the library that does this is called **Celery**. There is also a Celery crate for Rust. This approach is also utilized for multiple server setups. Considering this, we could be tempted to send tasks to the queue inside another task. However, we can see here that this approach can lock up our queue:

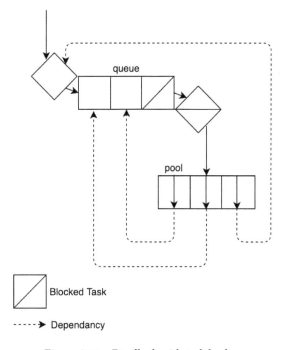

Figure 3.11 – Deadlock with task broker

In *Figure 3.11*, we can see that the tasks in the pool have sent tasks to the queue. However, they cannot complete until their dependencies have been executed. The thing is, they will never execute because the pool is full of tasks waiting for their dependency to complete and the pool is full, so they cannot be processed. The issue with this problem is that there are no errors raised with this—the pool will just hang. Deadlock is not the only problem that will arise without helpful warnings. Considering this, we must cover our last concept that we should be aware of before being creative: **race conditions**.

Race conditions

Race conditions occur when two or more threads access shared data that they both try to change. As we have noted when we were building and running our threads, they sometimes ran out of order. We can demonstrate this with a simple concept, as follows:

- If we were to have *thread one* calculate a price and write to a file and *thread two* to also calculate a price, read the price calculated from the *thread one* file, and add them together, there could be a chance that the price will not be written to the file before *thread two* reads it. What is even worse is that there could be an old price in the file. If this is the case, we will never know that the error occurred. The term *race conditions* is built upon the fact that both threads are racing to the data.

As a solution to race conditions, we can introduce **locks**. Locks can be utilized for stopping other threads from accessing certain things such as a file until your thread has finished with it. However, it has to be noted that these locks only work inside the process; therefore, other processes can access the file. Caching solutions such as Redis and general databases have already implemented these safeguards, and locks will not protect against the race condition described in this section. In my experience, when we get creative with thread concepts such as locks, it is usually a sign that we must take a step back and rethink our design.

Even an SQLite database file will manage our data race issues when reading and writing to a file, and if the data race condition described at the start of this section looks like it might happen, it is best to just not have them running at the same time at all. Sequential programming is safer and more useful.

Summary

In this chapter, we went through the basics of multiprocessing and multithreading. We then went through practical ways to utilize threads and processes. We then explored the Fibonacci sequence to explore how processes can speed up our computations. We also saw through the Fibonacci sequence that how we compute our problems is the biggest factor over threads and processes. Algorithms that scale exponentially should be avoided before reaching for multiprocessing for speed gains. We must remember that while it might be tempting to reach for more complex approaches to multiprocessing, this can lead to problems such as deadlock and data races. We kept our multiprocessing tight by keeping it contained within a processing pool. If we keep these principles in mind and keep all our multiprocessing contained to a pool, we will keep our hard-to-diagnose problems to a minimum. This does not mean that we should never be creative with multiprocessing but it is advised to do further reading on this field, as there are books entirely dedicated to concurrency (as noted in the *Further reading* section, with particular chapters to focus on). This is just an introduction to enable us to use concurrency in our Python packages if needed. In the next chapter, we will be building our own Python packages so that we can distribute our Python code across multiple projects and reuse code.

Questions

1. What is the difference between a process and a thread?

2. Why wouldn't multithreading speed up our Python Fibonacci sequence calculations?

3. Why is a multiprocessing pool used?

4. Our threads in Rust return `Result<i8, Box<dyn Any + Send>>`. What does this mean?

5. Why should we avoid using a recursion tree if we can?

6. Should you just spin up more processes when you need a faster runtime?

7. Why should you avoid complex multiprocessing if you can?

8. What does `join` do for our program in multithreading?

9. Why does `join` not return anything in a process?

Answers

1. Threads are lightweight and enable multithreading, where we can run multiple tasks that could have idle time. A process is more expensive, enabling us to run multiple CPU-heavy tasks at the same time. Processes do not share memory, while threads do.

2. Multithreading would not speed up our Fibonacci sequence calculations because calculating Fibonacci numbers is a CPU-heavy task that does not have any idle time; therefore, the threads would run sequentially in Python. However, we did demonstrate that Rust can run multiple threads at the same time, getting a significant speed increase.

3. Multiprocessing is expensive and the processes do not share memory, making the implementation potentially more complex. A processing pool keeps the multiprocessing part of a program to a minimum. This approach also enables us to easily control the different numbers of workers we need as they're all in one place, and we can also return all the outcomes in the same sequence as they are returned from the multiprocessing pool.

4. Our Rust thread could fail. If it doesn't, then it will return an integer. If it fails, it could return anything of any size, which is why it's on the heap. It also has the `Send` trait, which means that it can be sent across threads.

5. Recursion trees scale exponentially. Even if we are multithreading, our computation time will quickly scale, pushing our milliseconds into seconds once we've crossed a boundary.

6. No—as demonstrated in Amdahl's law, increasing the workers will give us some speedup, but we will have diminishing returns as the number of workers increases.

7. Complex multiprocessing/multithreading can introduce a range of silent errors such as deadlock and data races that can be hard to diagnose and solve.

8. `join` blocks the program until the thread has completed. It can also return the result of the thread if we overwrite Python's `join` function.

9. Processes do not share the same memory space, therefore they cannot be accessed. We can, however, access data from other processes by saving data to files for our main process to access or pipe data via `stdin` and `stdout`, as we did in our Rust multiprocessing example.

Further reading

- *Pan Wu (2020). Understanding Python Multithreading and Multiprocessing via Simulation:* `https://towardsdatascience.com/understanding-python-multithreading-and-multiprocessing-via-simulation-3f600dbbfe31`

- *Brian Troutwine (2018). Hands-On Concurrency with Rust*

- *Gabriele Lanaro and Quan Nguyen (2019). Learning Path Advanced Python Programming: Chapter 8 (Advanced Introduction to Concurrent and Parallel Programming)*

- *Andrew Johnson (2018). Hands-On Functional Programming in Rust: Chapter 8 (Implementing Concurrency)*

- *Rahul Sharma and Vesa Kaihlavirta (2018). Mastering Rust: Chapter 8 (Concurrency)*

Section 2: Fusing Rust with Python

Now that you are familiar with Rust, we can start utilizing it. Before we do this, we need to cover how to build Python packages that can be installed with pip. Once this has been done, we can build Python pip modules in Rust. This is where we can import our compiled Rust code into our Python code and run it with all the benefits of Rust in our Python application. We then go further into this by working with Python objects and using Python modules inside the Rust code.

This section comprises the following chapters:

- *Chapter 4, Building pip Modules in Python*
- *Chapter 5, Creating a Rust Interface for Our pip Module*
- *Chapter 6, Working with Python Objects in Rust*
- *Chapter 7, Using Python Modules in Rust*
- *Chapter 8, Structuring an End-to-End Python Module in Rust*

4

Building pip Modules in Python

Writing code to solve our problems is useful. However, writing code can become repetitive and time-consuming, especially when we are building applications. Applications usually require defining the steps that build the application. Packaging our code can help us reuse our code and share it with other developers. In this chapter, we will package Fibonacci code into a Python `pip` module that can be easily installed and has a command-line tool. We will also cover continuous integration processes that deploy our packages once a merge has been achieved to our `main` branch.

In this chapter, we will cover the following topics:

- Configuring setup tools for a Python `pip` module
- Packaging Python code in a `pip` module
- Configuring continuous integration

Technical requirements

We will need to have Python 3 installed. To get the most out of this chapter, we will also need to have a GitHub account, as we will be using GitHub to package our code, which can be accessed via this link: `https://github.com/maxwellflitton/flitton-fib-py`.

Git command-line tools are also needed in this chapter. These can be installed by following the instructions here: `https://git-scm.com/book/en/v2/Getting-Started-Installing-Git`. The chapter will also make use of a PyPI account. You will need to have your own PyPI account, which can be obtained for free with this link: `https://pypi.org/`.

The code for this chapter can be found via this link: `https://github.com/PacktPublishing/Speed-up-your-Python-with-Rust/tree/main/chapter_four`.

Configuring setup tools for a Python pip module

Setup tools in Python are how the code in our module is packaged and installed. They provide a set of commands and parameters for the system that is installing the code to process. To explore how this is done, we will package the Fibonacci numbers example introduced in the previous chapter. However, these calculations will be packaged in a `pip` module. To configure our setup tools, we are going to have to carry out the following steps:

1. Create a GitHub repository for our Python `pip` package.
2. Define basic parameters.
3. Define a `README` file.
4. Define a basic module structure.

Let's have a look at each of these steps in detail in the following subsections.

Creating a GitHub repository

Understandably, a seasoned developer can create a GitHub repository but for the sake of completeness, we will offer all the steps needed. If you can already create a GitHub repository, move on to the next subsection:

1. On the home URL of GitHub when we are logged in, we can create our repository by clicking on the **New** button, as shown here:

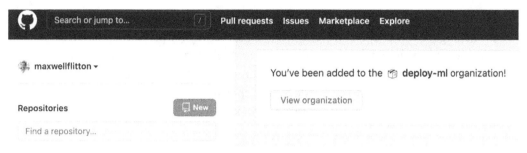

Figure 4.1 – How to create a new repository on GitHub

2. Once this is clicked, we can configure our new repository with the parameters
    shown next:

Create a new repository

A repository contains all project files, including the revision history. Already have a project repository
elsewhere? Import a repository.

Repository template
Start your repository with a template repository's contents.

No template ▾

Owner *   **Repository name** *

🐙 maxwellflitton ▾   /   flitton-fib-py   ✓

Great repository names are short and memorable. Need inspiration? How about legendary-carnival?

Description (optional)

This is a basic pip module on calculating Fibonacci numbers

⦿ 🖥 **Public**
        Anyone on the internet can see this repository. You choose who can commit.

◯ 🔒 **Private**
        You choose who can see and commit to this repository.

Initialize this repository with:
Skip this step if you're importing an existing repository.

☑ **Add a README file**
    This is where you can write a long description for your project. Learn more.

☑ **Add .gitignore**
    Choose which files not to track from a list of templates. Learn more.

    .gitignore template: **Python** ▾

☑ **Choose a license**
    A license tells others what they can and can't do with your code. Learn more.

    License: MIT License ▾

This will set ⑂ main as the default branch. Change the default name in your settings.

Create repository

Figure 4.2 – Parameters for our new GitHub repository

For this example, we have set the GitHub repository to **Public**; however, our `pip` packaging for this chapter will also work the same way for private repositories. We have also included a `.gitignore` file and selected it to be Python. This is to stop Python caching and for virtual environment files to be tracked by GitHub and uploaded when we upload our code to the repository. Now that we have our GitHub repository made, going to the repository will look like this:

Figure 4.3 – Our GitHub repository home page

We can see that our description is written in the README.md file. It also has to be noted that the README.md file is rendered. This happens in any directory of the repository. We can document what to do and how to use the code throughout the repository with a range of README.md files if we want.

3. Once this is done, we can download our repository with the command shown next:

```
git clone https://github.com/maxwellflitton/flitton-
    fib-py.git
```

Your URL will be different, as you have a different repository. The only thing left is to ensure that our development environment for our repository has a Python virtual environment.

4. This can be done by navigating to the root directory of the GitHub repository and then running the command shown here:

```
python3 -m venv venv
```

This creates a Python virtual environment under the `venv` directory in the root directory. We have to use the `venv` directory, as this is automatically included in the `.gitignore` file. However, there is nothing stopping us from calling it what we want, as long as we include it in the `.gitignore` file. However, `venv` is the convention, and using this will avoid confusion with other developers. Our environment is now fully set up.

5. To use our virtual environment in the terminal, we can activate it with the command shown next:

```
source venv/bin/activate
```

We can see that our command is prefixed with `(venv)`, meaning that it is active.

Defining the basic parameters

Now that our environment is fully functional, we are going to define the basic parameters when installing our Python `pip` module:

1. This is achieved by creating a `setup.py` file in the root of our repository. This will get run when another Python system installs our `pip` module. In our `setup.py` file, we import our setup tools with the code shown here:

```
from setuptools import find_packages, setup
```

We are going to use `setup` to define our parameters, and we are going to use `find_packages` to exclude tests.

2. Now that we have imported our setup tools, we can define our parameters in the same file with the code shown here:

```
setup(
    name="flitton_fib_py",
    version="0.0.1",
    author="Maxwell Flitton",
    author_email="maxwell@gmail.com",
    description="Calculates a Fibonacci number",
    long_description="A basic library that \
      calculates Fibonacci numbers",
    long_description_content_type="text/markdown",
    url="https://github.com/maxwellflitton/flitton- \
      fib-py",
```

```
    install_requires=[],
    packages=find_packages(exclude=("tests",)),
    classifiers=[
        "Development Status :: 4 - Beta",
        "Programming Language :: Python :: 3",
        "Operating System :: OS Independent",
    ],
    python_requires='>=3',
    tests_require=['pytest'],
)
```

There are a lot of parameters here. What we have done from the `name` field to `url` is essentially define the metadata around our `pip` module. The `classifiers` fields are also metadata around our module. The rest of the fields have the following effects:

- The `Install_requires` field is currently an empty list. This is because our module is not requiring any third-party modules right now. We will cover dependencies in *Managing dependencies* section.

- The `packages` field ensures that we exclude our `test` directory when we start building our tests for our module. While we will use tests to check our module and ensure standards, we do not need to install them when we are using our module as a third-party dependency.

- The `Python_requires` field ensures that the system installing our module has the correct version of Python installed.

- `tests_require` is a list of requirements when tests are running.

3. Now that we have defined our basic setup, we can upload our code with the following commands:

```
git add -A
git commit -m "adding setup to module"
git push origin main
```

What we have done here is add all of the new and changed files to our Git branch (which is `main`). We then committed our files with the `adding setup to module` message. We then pushed our code to the `main` branch, which means that we uploaded our changes onto the Git repository online. This is not the best way to manage our code iterations. We will go over different branches and how to manage them in the continuous integration section near the end of this chapter.

You may have noticed that `long_description` is a Markdown; however, trying to fit an entire Markdown into this field would end up dominating the `setup.py` file. It would essentially be a long string spanning multiple lines, with a few Python lines dispersed into it. We want our `setup.py` file to dictate the logic of setting up the module when it is being installed. We also want our long description of the module to be rendered by GitHub when we visit the GitHub repository directly online. Because of this, we will need to add some extra logic around defining our long description in the next section.

Defining a README file

Our long description is essentially the `README.md` file. If we fuse this with our `setup.py`, our `README.md` file will also render if we visit it on PyPI and it is uploaded to the PyPI server. This can be done by reading the `README.md` file into a string in the `setup.py` file and then plugging that string into our `long_description` field with the following code in the `setup.py` file:

```python
with open("README.md", "r") as fh:
    long_description = fh.read()


setup(
    name="flitton_fib_py",
    version="0.0.1",
    author="Maxwell Flitton",
    author_email="maxwell@gmail.com",
    description="Calculates a Fibonacci number",
    long_description=long_description,
    ...
```

The rest of the code after `...` is the same as before. With this, our basic module setup is complete. Now, all we need is to do is define a basic module to install and use, which is what we will do in the next step.

Defining a basic module

Defining a basic module takes the following structure:

```
├── LICENSE
├── README.md
├── flitton_fib_py
│   └── __init__.py
```

```
├── setup.py
└── venv
```

We house the actual code that the user will have in our `flitton_fib_py` directory. For now, we are just going to have a basic function that prints something out so that we can see if our `pip` package works. Here are the steps:

1. We do this by adding a basic `print` function in the `flitton_fib_py/__init__.py` file that has the following code:

    ```
    def say_hello() -> None:
        print("the Flitton Fibonacci module is saying hello")
    ```

    Once this is done, we can upload the code to the GitHub repository using the git commands described in the *Packaging Python code in a pip module* section. We should now see all the code of our module in the `main` branch. Considering this, we need to navigate to another directory that is not associated with our `git` repository.

2. We then unlink our virtual environment by typing the following command:

    ```
    deactivate
    ```

    We can then create a new virtual environment using the steps covered in the first section and activate it. Now, we are ready to install our package in our new virtual environment using `pip install` and check to see whether it works.

3. To use `pip install`, we point to the URL of the GitHub repository that our `pip` module is stored and define which branch it is. We do this by typing the following command, all in one line:

    ```
    pip install git+https://github.com/maxwellflitton/
        flitton-fib-py@main
    ```

    Your GitHub repository will have a different URL and you might have a different directory. Running this command will give us a range of printouts, stating that it is cloning the repository and installing it.

4. We then open up a Python terminal by typing in the following command:

    ```
    python
    ```

5. We now have an interactive terminal. We can check to see whether our module works by typing in the following commands:

```
>>> from flitton_fib_py import say_hello
>>> say_hello()
```

Once the last command is typed, we will get the following printout in the terminal:

```
the Flitton Fibonacci module is saying hello
```

There we have it – our Python package works! This works for both private and public GitHub repositories. Nothing is stopping us now from packaging private Python code to reuse on other private Python projects!

While this is a useful tool to package and install code on other computers with minimal setup, we have to be careful. When we are running the setup.py file, we are running the code as our root user. Therefore, we have to ensure that we trust what we are installing. Putting malicious code into the setup.py file is a vector of attack. We can run direct commands on the computer using the SubProcess object from the standard Python library. Make sure you trust the author of the code that you are installing with pip install.

This also highlights how vigilant you have to be when merely just running pip install. There are developers out there who will slightly change a package. For instance, a famous case was the requests package. This is a common, well-used package; however, for some time, there was an imitation package called request. They relied on people mistyping pip install and downloading the wrong package. This is known as **typosquatting**.

We have now packaged our Python code into a module. However, it is not a very useful module. This brings us to our next section, where we package our Fibonacci sequencing code.

Packaging Python code in a pip module

Now that we have our GitHub repository configured, we can start building out our Fibonacci code for our module. To achieve this, we must carry out the following steps:

1. Build our Fibonacci calculation code.
2. Create a command-line interface.
3. Test our Fibonacci calculation code with unit tests.

Let's now discuss each of these steps in detail.

Building our Fibonacci calculation code

When it comes to building our Fibonacci calculation code, we will have two functions – one that will calculate a Fibonacci number and another that will take a list of numbers and lean on the calculation function to return a list of calculated Fibonacci numbers. For this module, we will take a functional programming approach. This does not mean that we should have a functional programming approach every time we build a `pip` module. We are using functional programming because Fibonacci sequence calculations naturally flow well with a functional programming style.

Python is an object-orientated language, and problems that have multiple moving parts interrelating naturally flow well with object-orientated approaches. Our module structure will take the following form:

```
├── LICENSE
├── README.md
├── flitton_fib_py
│   ├── __init__.py
│   └── fib_calcs
│       ├── __init__.py
│       ├── fib_number.py
│       └── fib_numbers.py
├── setup.py
```

For this chapter, we will maintain a simple interface so that we can focus on the packaging of code in a `pip` module. Here are the steps:

1. First, we can build our Fibonacci number calculator in the `fib_number.py` file with the following code:

    ```python
    from typing import Optional


    def recurring_fibonacci_number(number: int) -> \
    Optional[int]:
        if number < 0:
            return None
        elif number <= 1:
            return number
        else:
            return recurring_fibonacci_number(number - 1) + \
                recurring_fibonacci_number(number - 2)
    ```

Here, it has to be noted that we are returning None when the input number is below zero. Technically, we should be throwing an error, but this is in place, for now, to demonstrate the effectiveness of a checking tool later on in our *Configuring continuous integration* section. As we know from the previous chapter, the preceding code will correctly calculate a Fibonacci number based on the input number.

2. Now that we have this function, we can depend on this to create a function that creates a list of Fibonacci numbers in our fib_numbers.py file with the following code:

```
from typing import List
from .fib_number import recurring_fibonacci_number


def calculate_numbers(numbers: List[int]) -> List[int]:
    return [recurring_fibonacci_number(number=i) \
        for i in numbers]
```

We are now ready to test our pip module again. We must push our code to the main branch on our repository again, uninstall our pip package in another virtual environment, and install again using pip install.

3. In our Python terminal with our new installed package, we can test our recurring_fibonacci_number function with the following console commands:

```
>>> from flitton_fib_py.fib_calcs.fib_number
    import recurring_fibonacci_number
>>> recurring_fibonacci_number(5)
5
>>> recurring_fibonacci_number(8)
21
```

Here, we can see that our Fibonacci function can be imported, and it works, calculating the correct Fibonacci numbers.

4. We can test our calculate_numbers with the following commands:

```
>>> from flitton_fib_py.fib_calcs.fib_numbers
    import calculate_numbers
>>> calculate_numbers([1, 2, 3, 4, 5, 6, 7])
[1, 1, 2, 3, 5, 8, 13]
```

Here, we can see that our `calculate_numbers` function also works. We have a fully functioning Fibonacci `pip` module. However, if we want to just calculate a Fibonacci number without coding a Python script, we should not have to go into a Python terminal. We can remedy this by building a command-line interface in the next step.

Creating a command-line interface

In order to build our command line function, our module can take the following structure:

```
├── LICENSE
├── README.md
├── flitton_fib_py
│   ├── __init__.py
│   ├── cmd
│   │   ├── __init__.py
│   │   └── fib_numb.py
│   └── fib_calcs
    . . .
```

To build our interface, we follow these steps:

1. We build the command-line interface in our `fib_numb.py` file with the following code:

```python
import argparse
from flitton_fib_py.fib_calcs.fib_number \
    import recurring_fibonacci_number


def fib_numb() -> None:
    parser = argparse.ArgumentParser(
        description='Calculate Fibonacci numbers')
    parser.add_argument('--number', action='store',
                        type=int, required=True,
                        help="Fibonacci number to be \
                        calculated")
    args = parser.parse_args()
    print(f"Your Fibonacci number is: " \
```

```
        f"{recurring_fibonacci_number \
    (number=args.number)}")
```

Here, we can see that we get the parameters passed in from the command line using the `argparse` module. Once we have obtained the parameters, we will then calculate the number and print it out. Now, for us to actually access it via the terminal, we have to point to it in the `setup.py` file at the root of the `pip` package by adding the following parameter in the `setup` object initialization:

```
entry_points={
    'console_scripts': [
        'fib-number = \
            flitton_fib_py.cmd.fib_numb:fib_numb',
    ],
},
```

Here, what we are doing is linking the `fib-number` console command with the function that we have just defined. After uninstalling our `pip` module in another virtual environment, uploading the changes to the `main` branch on our repository, and installing our new module using `pip install`, we will have our new module with the command-line tool that we have built.

2. Once it is installed, we just type in the following command:

    ```
    fib-number
    ```

We get the following output:

```
usage: fib-number [-h] --number NUMBER
fib-number: error: the following arguments are
required: --number
```

Here, we can see that the `argparse` module that we are using ensures that we provide the arguments needed. If we need help, we can get this by typing in the following command:

```
fib-number -h
```

This gives us the help printout, as shown here:

```
usage: fib-number [-h] --number NUMBER


Calculate Fibonacci numbers


optional arguments:
```

```
     -h, --help          show this help message and exit
     --number NUMBER  Fibonacci number to be calculated
```

We can see that we have the type and the help description of what it does.

3.   So, to calculate the Fibonacci number, we use the following command:

```
fib-number --number 20
```

This gives us the following printout:

```
Your Fibonacci number is: 6765
```

If we were to provide a string instead of a number for our argument, our program would refuse it, throwing an error.

Here we have it, we have a fully working command-line tool! This does not stop here. You can take this further. Nothing is stopping you from using subprocess from the standard library combined with other libraries, such as Docker, to build your own DevOps tools. You can automate whole workflows for yourself and the applications you make. However, if we are to rely more and more on our pip modules to do the repetitive heavy lifting, we can get into serious problems if the program introduces some bugs we need to know straight away. To do this, we need to start building unit tests for our module. These are covered in the next subsection.

Building unit tests

Unit tests are helpful for us to check and maintain quality control for our code. To build our unit tests, our module will have the following structure:

```
├── LICENSE
├── README.md
├── flitton_fib_py
      . . .
├── scripts
│     └── run_tests.sh
├── setup.py
├── tests
│     ├── __init__.py
│     └── flitton_fib_py
│           ├── __init__.py
│           └── fib_calcs
│                 ├── __init__.py
```

```
|                    ├── test_fib_number.py
|                    └── test_fib_numbers.py
```

We can see that we are mimicking the structure of the code in our module. This is important to keep track of our tests. If the module grows, then we will not get lost in our tests. If we need to chop out a directory or move it to another module, we can simply delete the appropriate directory or move it. It also has to be noted that we have built a Bash script to run our tests.

When it comes to writing our tests, it is usually best to code based on the chain of dependency. For instance, our files have the dependency chain depicted as follows:

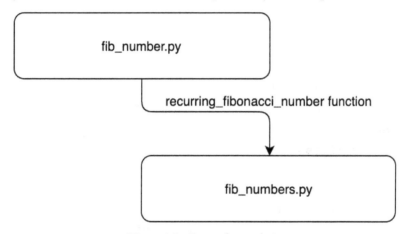

Figure 4.4 – Dependency chain

Considering our dependency chain, we should ideally write our tests for the fib_number.py file first and make sure that our recurring_fibonacci_number function works before writing tests that rely on the recurring_fibonacci_number function. Here are the steps to write our tests:

1. We first import what we need to test our code in our test_fib_number.py file via the following code:

    ```
    from unittest import main, TestCase

    from flitton_fib_py.fib_calcs.fib_number \
        import recurring_fibonacci_number
    ```

    The main function is to run all tests. We also rely on the TestCase class by writing our own test class that inherits TestCase. This gives our class extra class functions to aid us in testing outcomes.

2. We can write our own tests for a range of inputs with the following code:

```python
class RecurringFibNumberTest(TestCase):

    def test_zero(self):
        self.assertEqual(0,
            recurring_fibonacci_number(number=0)
        )
    def test_negative(self):
        self.assertEqual(
            None, recurring_fibonacci_number \
            (number=-1)
        )
    def test_one(self):
        self.assertEqual(1, \
            recurring_fibonacci_number(number=1))
    def test_two(self):
        self.assertEqual(1, \
            recurring_fibonacci_number(number=2))
    def test_twenty(self):
        self.assertEqual( \
        6765, recurring_fibonacci_number(number=20)
        )
```

Here, it has to be noted that each one of our functions has the `test_` prefix. This flags the function as a test function. This is also the case for the name of the file. All test files have the `test_` prefix to flag that the file houses tests. In our testing code, we can see that we have merely passed a range of inputs into the function that we are testing and asserted that the outcome is what we expect. If the assertions do not hold water, then we get an error and a failed result. Seeing as we are just testing the same function repeatedly, we can put all of the assertions into one test function. This is usually preferred if we are testing the whole object. We would essentially have one test function for each function that we are testing in the object.

3. Now that all our tests have been run, we can run the `unittest` `main` function if
    the `test_fib_number.py` file is run directly at the bottom of the `test_fib_`
    `number.py` file with the code shown next:

    ```
    if __name__ == "__main__":
        main()
    ```

4. We now have to set our `PYTHONPATH` variable to the directory of `flitton_`
    `fib_py`.

    Once this is done, we can run our `test_fib_number.py` file and get the console
    printout as shown:

    ```
    . . . . .
    ---------------------------------------------------------
    Ran 5 tests in 0.002s


    OK
    ```

    We can see that each test function is a test. The dots at the top are each test. If we
    were to change the `None` to a `1` in the second test, we would get the following
    printout:

    ```
    F. . . .
    =========================================================
    FAIL: test_negative (tests.flitton_fib_py.fib_calcs.
    test_fib_number.RecurringFibNumberTest)
    ---------------------------------------------------------
    Traceback (most recent call last):
      File "/Users/maxwellflitton/Documents/github/
      flitton-fib-py/tests/flitton_fib_py/fib_calcs/
      test_fib_number.py", line 15, in test_negative
        self.assertEqual(
    AssertionError: 1 != None
    ---------------------------------------------------------
    Ran 5 tests in 0.003s
    ```

    We can see that we now have an `F` in the test dots, and it highlights what test is
    failing and where it is failing.

5.   Now that we have built our base test, we can build our tests for the function that takes in a list of integers and returns a list of Fibonacci numbers. In our `test_fib_numbers.py` file, we import what we need with the following code:

```
from unittest import main, TestCase
from unittest.mock import patch


from flitton_fib_py.fib_calcs.fib_numbers \
    import calculate_numbers
```

Here, we can see that we are importing the function that we are testing and the same `main` and `TestCase`. However, it has to be noted that we have imported a `patch` function. This is because we have already tested our `recurring_fibonacci_number` function. The `patch` function enables us to insert a `MagicMock` object in place of our `recurring_fibonacci_number` function.

For our example, it can be argued that we do not need to patch anything. However, it is important to get an understanding of patching. **Patching** enables us to bypass expensive processes. For instance, if we are relying on a function that must make an API call, we should not have to make those API calls when testing. Instead, we can just patch the function. This also isolates the test. If a particular test is failing, we know that is something to do with the code that we are testing directly and not external code that it is depending on. It also speeds up the testing as we are not fully running code that we are depending on multiple times in different tests. We also get granularity because we use a `MagicMock` object; we can define the return values to anything we want during the test and log all calls to the `MagicMock` object.

The advantage here is that we might accidentally call the function we are depending on twice for some reason. However, if the function returns the same value twice, we will not know anything if we did not patch it. However, with patching, we can inspect the calls and throw errors if the behavior is not what we expect. We can also test a range of edge cases very quickly by merely changing the return value of the patches and rerunning the test.

With all this, it is understandable that we can get excited about patching. However, there are some downsides. If we do not update the patches' return values, the dependent code does not get the changes, and the testing does not remain accurate. This is why it is always sensible to have a mixture of approaches and run a functional test that runs the whole process without patching anything. With all this in mind, our patched unit test in the `tests/flitton_fib_by/fib_calcs/test_fib_numbers.py` file is carried out by the following code:

```python
class Test(TestCase):

    @patch("flitton_fib_py.fib_calcs.fib_numbers."
           "recurring_fibonacci_number")
    def test_calculate_numbers(self, mock_fib_calc):
        expected_outcome = [mock_fib_calc.return_value,
                            mock_fib_calc.return_value]
        self.assertEqual(expected_outcome,
                        calculate_numbers(numbers=[3, 4]))

        self.assertEqual(2,
            len(mock_fib_calc.call_args_list))
        self.assertEqual({'number': 3},
            mock_fib_calc.call_args_list[0][1])
        self.assertEqual({'number': 4},
            mock_fib_calc.call_args_list[1][1])
```

Here, we can see that we have used the patch as a decorator with a string that defines the path to the function that we are patching. We then pass the patched function through the test function under the `mock_fib_calc` parameter. We then state that we expect the outcome of the function that we are directly testing (`calculate_numbers`) to be a list of two return values of the patched function. We then pass two integers wrapped in a list into the `calculate_numbers` function and assert that this is going to be the same as our expected outcome. Once this is done, we assert that the `mock_fib_calc` was only called twice, and we inspect each of those calls, asserting that they are the numbers that we passed in, in the correct order. This has given us a lot of power to truly inspect our code. However, we are not done yet; we also must define the functional test to enable us to run our tests with the code here:

```python
    def test_functional(self):
        self.assertEqual([2, 3, 5],
```

```
        calculate_numbers(numbers=[3, 4, 5]))

if __name__ == "__main__":
    main()
```

For our module, all our unit tests are done. However, we do not want to go through manually running each file to see our tests. There will be times where we want to just see all the outcomes of the tests to see if there are any fails. To automate this, we can build a Bash script in the run_tests.sh file with the code here:

```
#!/usr/bin/env bash

SCRIPTPATH="$( cd "$(dirname "$0")" ; pwd -P )"
cd $SCRIPTPATH
cd ..

source venv/bin/activate
export PYTHONPATH="./flitton_fib_py"

python -m unittest discover
```

Here, we claim that the file is a Bash script with the first line. The first line is a shebang line and tells the computer running it what type of language it is. We then get the directory path of where this script is and assign it to the SCRIPTPATH variable. We then navigate to this directory, move out to the root of our module, activate our virtual environment, and then define our PYTHONPATH variable to be in our module with the Fibonacci number code. Now that everything is defined, to run our test we use the unittest command-line tool to run all the unit tests. Remember, all our tests have the test_ prefix in their filenames. Running this gives us the following printout:

```
.......
----------------------------------------
Ran 7 tests in 0.003s

OK
```

Here, we can see that we have seven tests running and they have all passed. We can see that we have started automating the test-running process. This is not where we should stop. As we move forward onto packaging and distributing our `pip` module, we should investigate automating the processes through continuous integration, which is what we explore in the next section. Right now, as it stands, if a user has access to our GitHub repository, we can install the code via `pip` and use it.

Configuring continuous integration

Our Python `pip` package is fully functioning. However, this is not the end. We will need to maintain the quality of the code and enable it to be constantly upgraded when we push new features to our module and refactor existing code. Continuous integration enables us to ensure that the tests pass and that the standard of quality is maintained. It also speeds up the deployment process, enabling us to push new iterations within a matter of minutes, enabling us to focus on the task at hand. It also reduces the risk of making a mistake.

As we know, the most mundane, repetitive tasks are the ones that are at the highest risk of a mistake occurring. This is just a fact of life. It's known that most car crashes happen within 5 minutes of the driver's home. This is because the drivers pay less attention and their brain switches off, relying on muscle memory. Deployment processes are the same. They are repetitive and do not require a lot of mental focus. As a result, after several times, we will start to rely on muscle memory and forget to check certain things, and make minor blunders when deploying our `pip` package. Continuous integration is a must to avoid mistakes and saves time in not only deployment but also in not having to correct the errors. To set up continuous integration, we are going to have to carry out the following steps:

1. Manually deploy onto PyPI.

2. Manage our dependencies.

3. Set up type checking for Python.

4. Set up and run tests and type checking with GitHub Actions.

5. Create automatic versioning for our `pip` package.

6. Deploy onto PyPI using GitHub Actions.

Let's have a look at each of these steps in detail in the following subsections.

Manually deploying onto PyPI

We now move on to our first step of manually deploying our GitHub repository onto PyPI. We have installed our `pip` package by directly pointing to the GitHub repository. However, if we are allowing everyone to access our module as it's open source, it is easier to upload our package onto PyPI. This will enable others to install using a simple command. Here are the steps:

1. First, we need to package our `pip` module before we upload it. This can be done with the following command:

    ```
    python setup.py sdist
    ```

    What this does is package our `pip` module in a `tar.gz` file, which gives us the following file outline:

    ```
    ├── LICENSE
    ├── README.md
    ├── dist
    │   └── flitton_fib_py-0.0.1.tar.gz
    ├── flitton_fib_py
        . . .
    ```

2. We can now see that the version is included in the filename. We are now ready to upload onto the PyPI server. To do this, we have to install `twine` with the following command:

    ```
    pip install twine
    ```

3. We are now able to upload the `tar.gz` file with the following command:

    ```
    twine upload dist/*
    ```

    This uploads all of the packages that we have created. During this process, the terminal will ask us for the PyPI username and password. It then uploads the package and tells us where we can find out the module on PyPI. If we visit this, we should get the view depicted in the following figure:

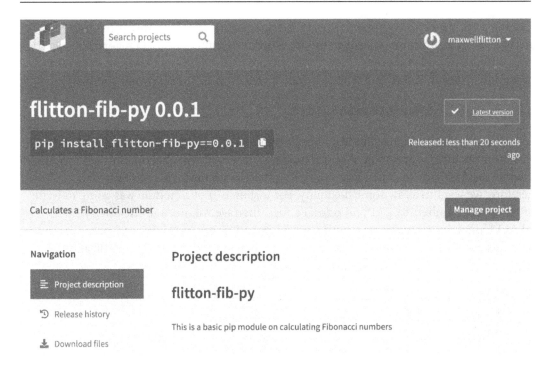

Figure 4.5 – PyPI view of our module

We can see that our README.md file is being directly rendered in the view in *Figure 4.5*. We can now directly install this with the pip install command depicted in the PyPI view. It must be noted that we now have a dependency. We need to manage these dependencies. We will cover this in the next step.

Managing dependencies

When it comes to dependencies, we must manage two types. For instance, our twine dependency helps us upload it onto PyPI. However, this is not needed for the pip package. Therefore, we need two different lists of dependencies – one for development and the other for actual use. We define the dependencies that we need for the development with the simple standard command stated here:

```
pip freeze > requirements.txt
```

What the `pip freeze` command gives us is a specific list of requirements that our current Python environment needs to install in order to run. `> requirements.txt` writes it to the `requirements.txt` file. If you are a new developer starting to develop our module, you can install all the requirements needed with the following command:

```
pip install -r requirements.txt
```

We can be strict here because nothing is depending on the development requirements apart from the direct development of our module. However, when it comes to our module, we know that it will be installed into multiple systems with multiple requirements. Therefore, we want to allow some flexibility. For instance, if our module was going to write our Fibonacci numbers to `yml` and `pickle` files, then we will need to use the `pyYAML` and `dill` modules to enable us to write our Fibonacci numbers to `yml` and `pickle` files. To do this, we alter our `install_requires` parameter in our `setup` initialization in the `setup.py` file with the code here:

```
install_requires=[
    "PyYAML>=4.1.2",
    "dill>=0.2.8"
],
```

It must be noted that these are not the latest packages. We must drop a few versions and allow our dependency to be equal to or above that version. This gives our users freedom when using our pip package in their systems. We also must copy and paste these requirements into our `requirements.txt` file to ensure that our development is consistent with the user experience of our `pip` module. Let's say that we are going to add an optional feature which is to start a small Flask server that locally serves an API that calculates Fibonacci numbers. Here, we can add an `install_requires` parameter in our `setup` initialization in the `setup.py` file with the following code:

```
extras_require={
    'server': ["Flask>=1.0.0"]
},
```

Now, if we upload our new code to either PyPI or our personal GitHub repository, we will have a different experience when installing our package. If we normally install it, we will see that our `pickle` and `yml` requirements automatically install if we run the install command, as shown here:

```
pip install flitton-fib-py[server]
```

It will actually install the server requirements. We can have as many requirements for the `server` profile as we want, and they will all be installed. Remember, our `extras_require` parameter is a dictionary, so we can define as many extra requirement profiles as we want. With this, we now have development requirements, essential `pip` module requirements, and optional `pip` module requirements. In the next step, we are now going to rely on a new development requirement to check types.

Setting up type checking for Python

At this point in the book, we have experienced the safety that Rust introduces. When types don't match up, the Rust compiler refuses to compile. However, with Python, we do not get this, as Python is an interpreted language. However, we can mimic this using the `mypy` module. The steps are as follows:

1. First, we can install the `mypy` module with the following command:

    ```
    pip install mypy
    ```

2. We can then type-check by using the `mypy` entry point with the code here:

    ```
    mypy flitton_fib_py
    ```

    Here, we are pointing to the main code for our Python module. We should get the console printout as follows:

    ```
    flitton_fib_py/fib_calcs/fib_number.py:16:
    error: Unsupported operand types for + ("int" and
    "None")
    flitton_fib_py/fib_calcs/fib_number.py:16:
    error: Unsupported operand types for + ("None" and
    "int")
    flitton_fib_py/fib_calcs/fib_number.py:16:
    error: Unsupported left operand type for + ("None")
    flitton_fib_py/fib_calcs/fib_number.py:16:
    note: Both left and right operands are unions
    flitton_fib_py/fib_calcs/fib_numbers.py:13:
    error: List comprehension has incompatible
    type List[Optional[int]]; expected List[int]
    ```

What mypy is doing is checking the consistency across all of our Python code! Like a Rust compiler, it has found an inconsistency. However, because this is Python, we can still run our Python code. While Python is memory-safe, the strong type-checking that Rust enforces is going to reduce the risk of incorrect variables being passed into the function in runtime. Now, we know that there is an inconsistency. The inconsistency is that our `recurring_fibonacci_number` function returns either None or int. However, our `calculate_numbers` function relies on the `recurring_fibonacci_number` function for the return value, but it returns a list of integers as opposed to returning a list of integers or None values.

3. We can constrict the return value to just an integer with the `recurring_fibonacci_number` function:

```
def recurring_fibonacci_number(number: int) -> int:
    if number < 0:
        raise ValueError(
        "Fibonacci has to be equal or above zero"
        )
    elif number <= 1:
        return number
    else:
        return recurring_fibonacci_number(number - 1) + \
            recurring_fibonacci_number(number - 2)
```

Here, we can see that we raise an error if the input number is below zero. It's not going to calculate anyway, so we might as well throw an error informing the user that there is an error as opposed to silently producing a None value.

If we run our mypy check, we get the following console printout:

```
Success: no issues found in 6 source files
```

Here, we can see that all our files were checked and that they have type consistency.

However, we might forget to run this type of checking every time we upload new code to the GitHub repository. In the next section, we will define GitHub Actions to automate our checking.

Setting up and running tests and type-checking with GitHub Actions

GitHub Actions run a series of computations that we can define in a `yml` file. We generally use GitHub Actions to automate processes that need to run every time. Workflow `yml` files are automatically detected by GitHub and run depending on what type of tags we give it. We can set up our GitHub Actions by following these steps:

1. For our tests and type-checking tags, we will define these in the `.github/workflows/run-tests.yml` file. In this file, we initially give a name for the workflow, and state that it fires when there is a push from one branch to another. This happens when a pull request is done as one branch is being pushed to another. This also reruns if we push more changes to our branch before merging the pull request. Our definitions are inserted at the top of the file with the following code:

    ```
    name: Run tests
    on: push
    ```

    Here, we can see that the workflow is called `Run tests`.

2. Next, we must define our jobs. We also must state that our job is a `shell` command. We then define what the operating system is. Once we have done this, we define the steps of the job. In our `steps` section, we then define the `uses`, which we will state are `actions` with the following code:

    ```
    jobs:
      run-shell-command:
        runs-on: ubuntu-latest
        steps:
          - uses: actions/checkout@v2
    ```

    If we did not define the `uses` step, then we would not be able to access files such as the requirements.

3. We are now ready to define the rest of the steps under the `steps` tag. These steps usually have a `name` and `run` tag. For us, we will be defining three steps:

    i.   The first one is to install the dependencies.

    ii. The second one is to run all the unit tests.

    iii. The third one is to run the type-checking with the code here:

```
- name: Install dependencies
  run: |
      python -m pip install -upgrade pip
      pip install -r requirements.txt
- name: run tests
  run: python -m unittest discover ./tests
- name: run type checking
  run: mypy flitton_fib_py
```

It must be noted that `run` is just a one-line terminal command. At one point, there is a | (pipe) value next to a `run` tag of the `Install dependencies` step. This pipe value simply allows us to write multiple lines of commands in one step. We must ensure that our `requirements.txt` file is updated with the `mypy` module. Once this is done, we can push this code to our GitHub repository and this GitHub action will run when we do pull requests. If you are familiar with GitHub and making pull requests, then you can move on to the next step. However, if you are not, then we can perform one now.

4. First, we have to pull a new branch from our `main` branch with the following command:

```
git checkout -b test
```

With this, we then have a branch called `test`. We can then make a change in our code.

5. To just trigger a GitHub action with a pull request, we can simply scar our code with a comment in any file, such as the one here:

```
# trigger build (14-6-2021)
```

You can write whatever, as it is just a comment if the code has changed. We then add and commit our changes to our **test** branch and push it to the GitHub repository. Once this is done, we can trigger a pull request by clicking on the **Pull requests** tab and selecting our **test** branch, as shown here:

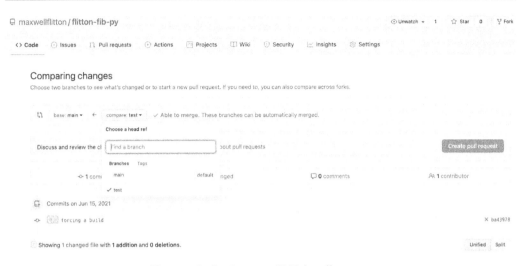

Figure 4.6 – Setting up a GitHub pull request

6. Once this is done, we can click on **Create pull request** to view it. Here, we will
    see all the GitHub Actions that get triggered and their status, as shown in the
    following figure:

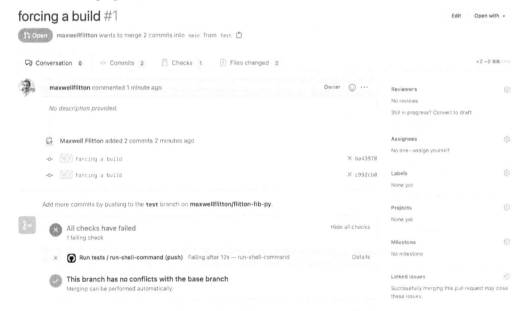

Figure 4.7 – View of the GitHub Actions status of pull requests

We can see that our tests have failed! If we click on **Details**, we can see that everything is working; it is just that we forgot to update our tests. If we remember, we changed our code to throw an error if we pass in a negative value into the Fibonacci calculation function, as shown next:

Figure 4.8 – View of the GitHub Actions execution details

7. We can change the test code to assert that an error is raised by the testing code in the tests/flitton_fib_py/fib_calcs/test_fib_number.py file with the following code:

```
def test_negative(self):
    with self.assertRaises(ValueError) as \
raised_error:
        recurring_fibonacci_number(number=-1)
    self.assertEqual(
        "Fibonacci has to be equal or above zero",
        str(raised_error.exception)
    )
```

Here, we can see that we assert that a value error is raised because we are running code that we expect to raise an error, and that the exception is what we expect of it. Pushing this to our GitHub repository will ensure that all the tests have passed. We can merge the pull request if we want the code to be merged into our main branch.

We have seen from this example that continuous integration is useful. It has picked up a change in the code that we might not have noted if we were doing everything manually.

Now that our tests run automatically, we need to automate keeping track of the version of our module to avoid making the same mistake we made with not updating our tests.

Create automatic versioning for our pip package

To automate the process of updating the version number, we are going to put several functions in the get_latest_version.py file in the root of our pip module. Following are the steps:

1. First, we need to import everything we need with the following code:

    ```
    import os
    import pathlib
    from typing import Tuple, List, Union
    import requests
    ```

    We are going to use os and pathlib to manage writing the latest version to a file. We are also going to use the requests module to call PyPI to get the latest version that is currently available to the public.

2. To do this, we can create a function that will get the metadata of our module from PyPI and return the version with the following code:

    ```
    def get_latest_version_number() -> str:
        req = requests.get(
        "https://pypi.org/pypi/flitton-fib-py/json")
        return req.json()["info"]["version"]
    ```

3. This is just a simple web request. Once we have done this, we are going to want to unpack this string into a tuple of integers with the function defined next:

    ```
    def unpack_version_number(version_string: str) \
        -> Tuple[int, int, int]:
        version_buffer: List[str] = \
          version_string.split(".")
        return int(version_buffer[0]),\
            int(version_buffer[1]),int(version_buffer[2])
    ```

Here, we can see that this is a simple split via bullet points. We then convert them to integers and pack them into a tuple to be returned.

4. Now that we have got our version number, we need to increase this by one with the function defined next:

```
def increase_version_number(version_buffer: \
Union[Tuple[int, int, int], List[int]]) -> List[int]:
    first: int = version_buffer[0]
    second: int = version_buffer[1]
    third: int = version_buffer[2]

    third += 1
    if third >= 10:
        third = 0
        second += 1
        if second >= 10:
            second = 0
            first += 1

    return [first, second, third]
```

Here, we can see that if one of the integers is equal or greater than 10, we set it back to 0 and increase the next number by 1. The only one that does not get sent to 0 is the furthest number to the left. This will just keep going up.

5. Now that we have increased our number by 1, we will need to pack the integer into a string, with the function defined next:

```
def pack_version_number(
    version_buffer: Union[Tuple[int, int, int],
    List[int]]) -> str:
    return f"{version_buffer[0]}.{version_buffer[1]} \
    .{version_buffer[2]}"
```

6. Once we have packed this into a string, we will have to write the version to a file. This can be done with the function defined next:

```
def write_version_to_file(version_number: str) -> \
None:
    version_file_path: str = str( \
```

```
        pathlib.Path(__file__).parent.absolute()) + \
        "/flitton_fib_py/version.py"

    if os.path.exists(version_file_path):
        os.remove(version_file_path)

    with open(version_file_path, "w") as f:
        f.write(f"VERSION='{version_number}'")
```

Here, we can see that we ensure that the path is going to be at the root of our module. We then delete the version file if it already exists, as it will already be out of date.

7. We then write our updated version number to the file with the following code:

```
if __name__ == "__main__":
    write_version_to_file(
        version_number=pack_version_number(
            version_buffer=increase_version_number(
                version_buffer=unpack_version_number(
                version_string=get_latest_version_number()
                )
            )
        )
    )
```

This ensures that if we run the file directly, we will get the updated version written to a file.

8. Now, in our `setup.py` file at the root of our module, we must read the version file and define it for our version parameter in the `setup` initialization. For that, we first import `pathlib` into our file and read the version file with this code:

```
import pathlib

with open(str(pathlib.Path(__file__).parent.absolute()) +
        "/flitton_fib_py/version.py", "r") as fh:
    version = fh.read().split("=")[1].replace("'", "")
```

9. We then set the `version` parameter with the read value with the following code:

```
setup(
    name="flitton_fib_py",
    version=version,
    . . .
```

We now have our version update process fully automated; we must plug this into our GitHub Actions, so we automatically run the update process and push to PyPI when merging with our `main` branch.

Deploying onto PyPI using GitHub Actions

To enable our GitHub actions to push to PyPI, we need to follow these steps:

1. First, we store the username and password for our PyPI account in the **Secrets** section of our GitHub repository. This can be done by clicking on the **Settings** tab and then the **Secrets** tab on the left sidebar, as shown here:

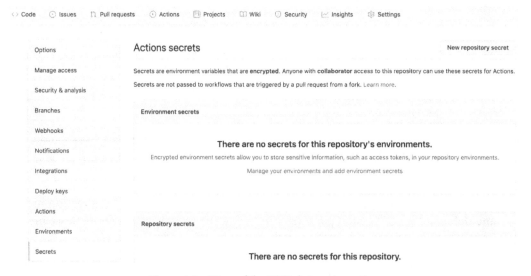

Figure 4.9 – View of the GitHub Secrets section

2. On the top right of the view in *Figure 4.9* is **New repository secret**. If we click this, we will get the following screen:

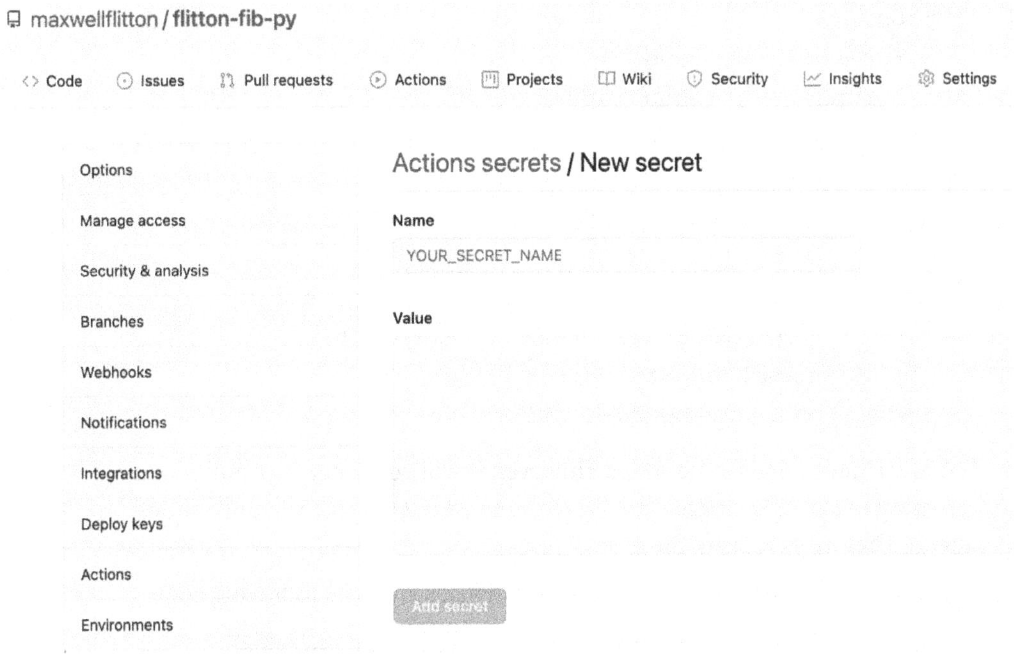

Figure 4.10 – View of the GitHub secret creation section

Here, we can create a secret for our PyPI password and another secret for our PyPI username.

Now that we have our secrets defined, we can build our GitHub action in the `.github/workflows/publish-package.yml` file:

1. First, we need to ensure that we publish our package only when we have merged a branch with the `main` branch. To do this, we need to ensure that our action only executes when there's a pull request when it's closed, and the branch being pointed out is `main` with the following code:

```
name: Publish Python 🐍 distributions 📦 to PyPI

on:
  pull_request:
    types: [closed]
    branches:
      - main
```

2. Once this is done, we can define the basic jobs of installing the dependencies and updating the package version with `jobs`, defined in the following code:

```
jobs:
  run-shell-command:
    runs-on: ubuntu-latest
    steps:
      - uses: actions/checkout@v2
      - name: Install dependencies
        run: |
          python -m pip install --upgrade pip
          pip install -r requirements.txt
      - name: update version
        run: python get_latest_version.py
```

What we have done is fine so far. However, it will run when any pull request pointing to `main` is closed. Therefore, we must ensure that the pull request has merged before executing the step.

3. For the next section, we install the dependencies with the following code:

```
      - name: install deployment dependancies
        if: github.event.pull_request.merged == true
        run: |
          pip install twine
          pip install pexpect
```

4. We can see that our conditional statements are straightforward. We then run the `setup.py` file for it to produce our distribution with the following step:

```
      - name: package module
        if: github.event.pull_request.merged == true
        run: python setup.py sdist
```

5. Now that we have defined all the steps needed to prepare our package, we can upload our package using `twine` with the following code:

```
      - name: deploy to pypi
        if: github.event.pull_request.merged == true
        env:
          TWINE_USERNAME: ${{ secrets.TWINE_USERNAME
```

```
                }}
        TWINE_PASSWORD: ${{ secrets.TWINE_PASSWORD
                }}
    run: |
        twine upload dist/*
```

Here, we can see that we have automated the deployment of our module to PyPI using GitHub Actions.

Summary

In this chapter, we have managed to build a fully fledged `pip` Python module that has continuous integration. We initially set up a GitHub repository and created a virtual environment. This is an essential skill for most Python projects, and you should be using GitHub repositories and virtual environments even if your project is not a `pip` module. You will be able to share your project and work with other team members. We then defined our `setup.py` file so our code could be installed via `pip`. Even if our GitHub repository is private, people who have access to the GitHub repository could freely install our code. This gives us even more power when it comes to distributing our code.

When we have an interface defined, our users do not need to know much about our code, just how to use the interface. This also enables us to prevent repeated code. For instance, if we build a user data model with a database driver, we can package it as a `pip` module and use this in multiple web applications. All we need to do is change the data model in the `pip` module and make a new release, and then all web applications can use the updated version if they wish.

Once our code was packaged, we rebuilt our Fibonacci code in our `pip` module, and it worked. We then went further, building entry points that enabled us to define our own command-line tools. This makes our code packaging even more powerful, as the user doesn't even have to import and code the module; they can just call the command-line argument! With this, we can build development tools to speed up our development by automating tasks with these entry points. We then built basic unit tests to ensure that the quality of our code was maintained. We then locked in these good standards with automation pipelines using GitHub Actions. We introduced type-checking with `mypy` alongside our unit-testing pipeline. We don't have to stop here. For instance, the Python script that we coded that increased the version number by one could be built in its own `pip` repository with a command-line interface. With this, we could install the module using `pip install` in our GitHub Actions and run the commands. Now, with this code packaging, you can build your own tools and add them to your belt, reducing the amount of repetition in your daily coding as time goes on.

In the next chapter, we cover what we have done in this chapter in Rust. Considering this, we harness the safety and speed of Rust, with the flexibility of `pip` packaging. Utilizing this will level up your skills as a Python toolmaker, making you invaluable to your team.

Questions

1. How would you perform an installation with `pip install` of our GitHub repository on the `test` branch?

2. Can other developers who do not have access to your GitHub repository install your `pip` package if you upload it to PyPI?

3. What is the difference between development dependencies and package dependencies?

4. `mypy` ensures the consistency of types when it comes to our Python code. How is this different from type-checking in Rust?

5. Why should we automate boring repetitive tasks?

Answers

1. `pip install git+https://github.com/maxwellflitton/flitton-fib-py@test`

2. Yes, they can download it despite not having access to your GitHub repository. If we think about it, we package our `pip` module in a file and then upload it to the PyPI server. Downloading our package from the PyPI server is not connected to our GitHub repository.

3. Development dependencies are specific dependencies defined in the `requirements.txt` file. This ensures that developers can work on the `pip` package. Package requirements are a little more relaxed and defined in the `setup.py` file. These get installed when the user installs our package. Package requirements are to enable the `pip` package to be used.

4. Rust does the type-checking when it is compiling and fails to compile if the types are inconsistent. Because of this, we cannot run it. Python, however, is an interpreted language. Because of this, we can still run it with the potential errors.

5. Repetitive tasks are easy to automate, so the effort invested is not excessive. Also, repetitive tasks have a higher risk of producing errors. Automating these tasks reduces the number of errors we could make.

Further reading

- *Python Organisation (2021) Packaging code*: `https://packaging.python.org/guides/distributing-packages-using-setuptools/`

- *GitHub Organisation (2021) GitHub Actions*: `https://docs.github.com/en/actions`

5

Creating a Rust Interface for Our pip Module

In *Chapter 4*, *Building pip Modules in Python*, we built a `pip` module in Python. Now, we will build the same `pip` module in Rust and manage the interface. Some people might prefer Python for some tasks; others will state that Rust is better. In this chapter, we will simply utilize both as and when we want. To achieve this, we will build a `pip` module in Rust that can be installed and directly imported into our Python code. We will also build Python entry points that talk directly to our compiled Rust code, and Python adapters/interfaces to make the user experience of our module easy, safe, and locked down with **user interfaces** (**UIs**) that have all features that we want our user to use.

In this chapter, we will cover the following topics:

- Packaging Rust with `pip`
- Building a Rust interface with the `pyO3` crate
- Building tests for our Rust package
- Comparing speed with Python, Rust, and Numba

Covering these topics will enable us to build Rust modules and use them in our Python systems. This is a major advantage as a Python developer; you can use faster, safer, and less resource-intensive code seamlessly in your Python programs.

Technical requirements

We will need to have **Python 3** installed. To get the most out of this chapter, we will also need to have a GitHub account, as we will be using GitHub to package our code, which can be accessed via this link: `https://github.com/maxwellflitton/flitton-fib-rs`.

The code for this chapter can be found at `https://github.com/PacktPublishing/Speed-up-your-Python-with-Rust/tree/main/chapter_five`.

Packaging Rust with pip

In this section, we will be setting up our `pip` package so that it can utilize Rust code. This will enable us to use Python setup tools to import our Rust `pip` package, compile it for our system, and use it within our Python code. For this chapter, we are essentially building the same Fibonacci module that we built in *Chapter 4, Building pip Modules in Python*. It is advised to create another GitHub repository for our Rust module; however, nothing is stopping you from refactoring your existing Python `pip` module. To build our Rust `pip` module, we are going to have to carry out the following steps:

1. Define `gitignore` and `Cargo` for our package.
2. Configure a Python setup process for our package.
3. Create a Rust library for our package.

Define gitignore and Cargo for our package

To get started, we must make sure that our Git does not track files that we do not want to upload and that our `Cargo` has the right dependencies with *step 1*.

1. First, we can start with `gitignore`. If you are choosing to use the same GitHub repository as the one that we defined in the previous chapter, then all the files for Python are already defined in the `.gitignore` file at the root of the GitHub repository. If not, then when you are creating your new GitHub repository, we have to select the Python template in the `Add .gitignore` section. Either way, once we have the Python `gitignore` template in our `.gitignore` file, we must add our `gitignore` requirements for the Rust part of our package. To do this, we add the following code in the `.gitignore` file:

    ```
    /target/
    ```

    Yes, that is it for our Rust code. That is a lot less than the Python files that we need to ignore.

2. Now that we have defined `gitignore`, we can move on to defining our `Cargo.toml` file in the root of our package, initially defining the metadata of our package with the following code:

    ```
    [package]
    name = "flitton_fib_rs"
    version = "0.1.0"
    authors = ["Maxwell Flitton
        <maxwellflitton@gmail.com>"]
    edition = "2018"
    ```

    This is nothing new; all we are doing here is defining the name and generic information of our package.

3. We then go on to define the dependencies with the following code:

    ```
    [dependencies]

    [dependencies.pyo3]
    version = "0.13.2"
    features = ["extension-module"]
    ```

We can see that we have not defined any dependencies in the dependencies section. We will be depending on the pyo3 crate to enable our Rust code to interact with our Python code. We declare the latest version of the crate at the point of writing this book, and the fact that we want to enable the extension-module feature because we will be using pyo3 to make our Rust module.

4. We then define our library data with the following code:

```
[lib]
name = "flitton_fib_rs"
crate-type = ["cdylib"]
```

It must be noted that we have defined a crate-type variable. Crate types provide information to the compiler on how to link Rust crates together. This can either be static or dynamic. For instance, if we were to define the crate-type variable as bin, this would compile our Rust code as a runnable executable. The main file would have to be present in our module, as this would be the entry point. We could also define the crate-type variable as lib, which compiles it as a library that can be used by other Rust programs. We can go further with this, defining either a static or dynamic library. Defining the crate-type variable as cdylib tells the compiler that we want a dynamic system library to be loaded by another language. If we do not put this in, we will not be able to compile our code when installing our library via pip. Our library should be able to compile for Linux and Windows. However, we require some link arguments to ensure that our library also works on macOS.

5. In order to do this, we need to define the configuration in the .cargo/config file:

```
[target.x86_64-apple-darwin]
rustflags = [
    "-C", "link-arg=-undefined",
    "-C", "link-arg=dynamic_lookup",
]
[target.aarch64-apple-darwin]
rustflags = [
    "-C", "link-arg=-undefined",
    "-C", "link-arg=dynamic_lookup",
]
```

With this, we have defined all that we need for our Rust library. Now, we move on to the next step, configuring the Python part of our module.

Configuring the Python setup process for our package

When it comes to setting up the Python section, we will be defining this in the `setup.py` file at the root of our module. Initially, we are going to import all the requirements that we need with the following code:

```
#!/usr/bin/env python
from setuptools import dist
dist.Distribution().fetch_build_eggs(['setuptools_rust'])
from setuptools import setup
from setuptools_rust import Binding, RustExtension
```

We are going to use the `setuptools_rust` module for managing our Rust code. However, we cannot be sure that the user will have installed `setuptools_rust` and we need it for running our setup code. Because of this, we cannot rely on the requirements list, as installing the requirements happens after we have imported `setuptools_rust`. To get around this, we use the `dist` module to get the required `setuptools_rust` module for this script. The user does not permanently install `setuptools_rust` but uses it for the script. Now that this is done, we can define our setup with the following code:

```
setup(
    name="flitton-fib-rs",
    version="0.1",
    rust_extensions=[RustExtension(
        ".flitton_fib_rs.flitton_fib_rs",
        path="Cargo.toml", binding=Binding.PyO3)],
    packages=["flitton_fib_rs"],
    classifiers=[
            "License :: OSI Approved :: MIT License",
            "Development Status :: 3 - Alpha",
            "Intended Audience :: Developers",
            "Programming Language :: Python",
            "Programming Language :: Rust",
            "Operating System :: POSIX",
            "Operating System :: MacOS :: MacOS X",
```

```
        ],
    zip_safe=False,
)
```

Here, we can see that we define the metadata of the module as we did in the previous chapter. We can also see that we define a `rust_extensions` parameter, pointing to the actual Rust module that we will define in a Rust file, as we can see in the following figure:

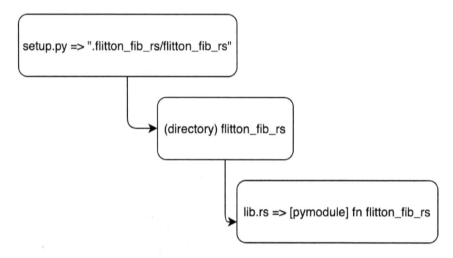

Figure 5.1 – Our module flow for setup

We also point to our `Cargo.toml` file, as we will have to compile other Rust crates that are in our dependencies when we are installing our Rust module. We also must state that our module is not zipped safely. This is also standard for C modules. Now that we have done all the setup configurations, we can now move on to the next step of building our basic Rust module that will get us installing Rust code using `pip install` that we can use in our Python code:

1. For our Rust code, we have to initially import all of the `pyo3` requirements in the `src/lib.rs` file with the following code:

    ```
    use pyo3::prelude::*;
    use pyo3::wrap_pyfunction;
    ```

    What this does is enable our Rust code to utilize all the macros that the `pyo3` crate has. We will also be wrapping the Rust functions into the module.

2. We then define a basic *hello world* function with the following code:

```
#[pyfunction]
fn say_hello() {
    println!("saying hello from Rust!");
}
```

We can see that we have applied a Python function macro from `pyo3` to the `say_hello` function.

3. Now that we have the function, we can define our module in the same file with the following code:

```
#[pymodule]
fn flitton_fib_rs(_py: Python, m: &PyModule) -> \
    PyResult<()> {
        m.add_wrapped(wrap_pyfunction!(say_hello));
        Ok(())
    }
```

Here, we can see that we have defined the module as `flitton_fib_rs`. This will have to be imported as `flitton_fib_rs` when using it. We then use the `pymodule` macro. This function is loading the module. We must define a result at the end. Seeing as we do not have any complex logic, we will define the end result as an `Ok` result. We don't need to do anything to Python; however, we add our wrapped `say_hello` function to our module. The `wrap_pyfunction` macro essentially takes a Python instance and returns a Python function. Now that we have our Rust code defined, we must build our Python entry point.

4. This is fairly simple; all we have to do is import our functions from the Rust module in the `flitton_fib_rs/__init__.py` file with the following code:

```
from .flitton_fib_rs import *
```

We will go through how this works later in this chapter, as we will be installing this package and running it.

Installing our Rust library for our package

Right now, we have everything we need to deploy our package and install it via `pip`. Considering this, we upload our package to our GitHub repository, which is covered in the *Configuring setup tools for a Python pip module* section of the *Chapter 4, Building pip Modules in Python*.

Once we have done this, we can install our `pip` package with the following command, all in one line:

```
pip install git+https://github.com/maxwellflitton/flitton-
    fib-rs@main
```

The URL to your GitHub repository might be different. When this is being installed, the process will hang for a while. The result should give the following printout:

```
Collecting git+https://github.com/maxwellflitton
/flitton-fib-rs@main
Cloning https://github.com/maxwellflitton/
flitton-fib-rs (to revision main) to /private
/var/folders/8n/
7295fgp11dncqv9n0sk6j_cw0000gn/T/pip-req-build-kcmv4ldt
Running command git clone -q https:
//github.com/maxwellflitton/flitton-fib-rs
/private/var/folders/8n
/7295fgp11dncqv9n0sk6j_cw0000gn/T/pip-req-build-kcmv4ldt
Installing collected packages: flitton-fib-rs
  Running setup.py install for flitton-fib-rs ... done
Successfully installed flitton-fib-rs-0.1
```

This is because we are compiling the package based on our system. Here, we can see that we collect the code from the `main` branch of the repository and run the `setup.py` file. What we have essentially done is compile the Rust code into a binary file and put it next to our `__init__.py` entry point file with the following file layout:

```
├── flitton_fib_rs
|    ├── __init__.py
|    └── flitton_fib_rs.cpython-38-darwin.so
```

This is why our `from .flitton_fib_rs import *` code works in the entry point.

Now that this is all installed in our Python packages, we can run our Python console and type in the following commands:

```
>>> from flitton_fib_rs import say_hello
>>> say_hello()
saying hello from Rust!
```

Here we have it! We have got Rust working with Python and we have managed to package our Rust code as a `pip` module. This is a complete game changer. We can now utilize Rust code without having to rewrite our Python systems. However, we only have one file in Rust code. We need to learn how to build bigger Rust systems if we want to fully take advantage of our ability to fuse Rust with Python.

Building a Rust interface with the pyO3 crate

Building an interface does not just mean adding more functions to our module in Rust and wrapping them. In a sense, we do have to do some of this; however, exploring how to import them from other Rust files is important. We also must explore and understand the relationship that we can have between Rust and Python when we are building up our module. To achieve this, we will carry out these steps:

1. Build our Fibonacci module in our Rust package.
2. Create command-line tools for our package.
3. Create adapters for our package.

With *step one*, we can just build out our module with Rust code. *Steps two* and *three* are more Python-focused, wrapping our Rust code in Python code to ease the interaction of our Rust module with external Python code. In *Chapter 6*, *Working with Python Objects in Rust*, we will interact directly with Python objects in our Rust code. With all this in mind, let's our Python interface by initially building our Fibonacci code in Rust with *step one*.

Building our Fibonacci Rust code

In this step, we are going to build our Fibonacci module, spanning multiple Rust files. To achieve this, the file structure of our module takes the following form:

```
├── Cargo.toml
├── README.md
├── flitton_fib_rs
│   ├── __init__.py
├── setup.py
├── src
│   ├── fib_calcs
│   │   ├── fib_number.rs
│   │   ├── fib_numbers.rs
```

```
|     |    └── mod.rs
|     ├── lib.rs
```

Here, we can see that we have added our Fibonacci code under the `src/fib_calcs` directory, as we remember that `fib_numbers.rs` relies on `fib_number.rs`.

Now, let's follow these steps:

1. We can initially define our Fibonacci number function in the `fib_number.rs` file with the following code:

```
use pyo3::prelude::pyfunction;


#[pyfunction]
pub fn fibonacci_number(n: i32) -> u64 {
    if n < 0 {
        panic!("{} is negative!", n);
    }
    match n {
        0       => panic!("zero is not a right \
                    argument to fibonacci_number!"),
        1 | 2 => 1,
        _       => fibonacci_number(n - 1) +
                    fibonacci_number(n - 2)
    }
}
```

Here, we can see that we have imported the `pyfunction` macro to apply to our function. By now, we are familiar with calculating a Fibonacci number; however, unlike previous examples, it must be noted that we have removed the `if the input Fibonacci number to be calculated is 3` match statement. This is because that match statement significantly speeds up the code, and we want a fair speed comparison for the final section of this chapter.

2. Now that we have defined our Fibonacci number function, we can define our `fibonacci_numbers` function in the `fib_numbers.rs` file with the following code:

```
use std::vec::Vec;
use pyo3::prelude::pyfunction;
use super::fib_number::fibonacci_number;
```

```
#[pyfunction]
pub fn fibonacci_numbers(numbers: Vec<i32>) -> \
   Vec<u64> {
      let mut vec: Vec<u64> = Vec::new();

      for n in numbers.iter() {
         vec.push(fibonacci_number(*n));
      }
      return vec
}
```

Here, we can see that we accepted a vector of integers, looped through them, and appended them to an empty vector, returning the vector with all the calculated Fibonacci numbers. Here, we have imported the fibonacci_number function.

3. However, we remember that we will not be able to import it, and neither of these functions will be available outside of the immediate directory if we do not define them in the src/mod.rs file with the following code:

```
pub mod fib_number;
pub mod fib_numbers;
```

4. Now that we have defined both of our functions and declared them in our src/mod.rs file, we are now able to import them into our lib.rs file. We do this by initially declaring the fib_calcs module, and then importing the functions with the following code:

```
mod fib_calcs;

use fib_calcs::fib_number::__pyo3_get_function \
   _fibonacci_number;
use fib_calcs::fib_numbers::__pyo3_get_function \
   _fibonacci_numbers;
pub mod fib_numbers;
```

Here, it must be noted that our functions have the prefix of __pyo3_get_function_. This enables us to retain the macros applied to the functions. If we just directly import the functions, we will not be able to add them to the module, which will result in compilation errors when we are installing our package.

5. Now that our functions are imported and ready, we can import-wrap them and add them to the module with the following code:

```
#[pymodule]
fn flitton_fib_rs(_py: Python, m: &PyModule) -> \
    PyResult<()> {
        m.add_wrapped(wrap_pyfunction!(say_hello));
        m.add_wrapped(wrap_pyfunction!(fibonacci_number));
        m.add_wrapped(wrap_pyfunction!(fibonacci_numbers));
        Ok(())
}
```

6. Now that we have built our modules, we can test them. We do this by uploading our changes to the GitHub repository, and using pip uninstall to uninstall our pip module and pip install to install our new package. Once our new package is installed, we can import and use our new functions in the Python terminal as follows:

```
>>> from flitton_fib_rs import fibonacci_number,
fibonacci_numbers
>>> fibonacci_number(20)
6765
>>> fibonacci_numbers([20, 21, 22])
[6765, 10946, 17711]
>>>
```

Here, we can see that we can import and use the Fibonacci numbers that we have coded in Rust that span multiple files! We are now at the stage where nothing is stopping us from building our own Rust Python pip packages. If you have a specific problem in mind to solve in Rust, such as an expensive computation that your Python program is struggling to calculate, nothing is stopping you from solving that now.

Now that we have gone to the trouble of packaging our Python package written in Rust, we can further utilize our package with command-line functionality. Packages installed with pip are convenient, powerful tools for command-line functionality. In the next section, we will access the Rust code in our package directly from the command line.

Creating command-line tools for our package

You may have noticed that to use our Fibonacci functions, we must start a Python console, import the functions, and use them. This is not very efficient if we just want to calculate a Fibonacci number in the console. We can remove these unnecessary procedures needed for calculating a Fibonacci number in the terminal by defining our entry points.

Considering that we define our command-line entry points in the setup.py file, it makes sense to define our entry point in a Python file that acts as a wrapper to our Rust function (as we still want the speed benefits of Rust), as shown in the following figure:

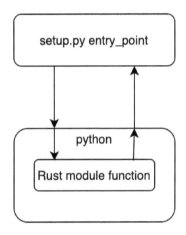

Figure 5.2 – Flow for module entry point

This wrapper can be done by importing argparse and the fibonacci_number function we made in the Rust module to create a simple Python function that gets user input and then passes it into the Rust function, printing out the result. We can achieve this by carrying out the following steps:

1. We can build the Python function that collects the arguments and calls the Rust code by adding the following code to the flitton_fib_rs/fib_number_command.py file we create:

```
import argparse
from .flitton_fib_rs import fibonacci_number


def fib_number_command() -> None:
    parser = argparse.ArgumentParser(
        description='Calculate Fibonacci numbers')
    parser.add_argument('--number', action='store', \
        type=int, required=True,help="Fibonacci \
```

```
        number to becalculated")
    args = parser.parse_args()
    print(f"Your Fibonacci number is: "
        f"{fibonacci_number(n=args.number)}")
}
```

We must remember that when our Rust binary is compiled, it will be in the flitton_fib_rs directory, right next to the file we just created.

2. Next, we define the entry point in the setup.py file. Now that we have our function, we can point to this in the setup.py file by declaring the path to this file and function for the entry_points parameter with the following code:

```
entry_points={
    'console_scripts': [
        'fib-number = flitton_fib_rs.'
        'fib_number_command:'
        'fib_number_command',
    ],
},
```

3. Once this is done, we have fully plumbed up the Python entry point in our package. Finally, we can test our command line by passing in arguments to the entry point. Now, if we update our GitHub repository and reinstall our package in the Python environment, we can test our command line by typing in the following command:

```
fib-number --number 20
```

This will give us the following output:

```
Your Fibonacci number is: 6765
```

We can see that our command-line tools work. Now we are at the stage where we have replicated the same functionality as our Python pip package previously in *Chapter 4, Building pip Modules in Python*. However, we must go further now. We are fusing two different languages in our package. To gain full command of our pip package, we need to explore how to command and refine the interaction between Rust and Python.

In our next step, we will build adapters that enable us to do this.

Creating adapters for our package

Before we try and build adapter interfaces, we need to understand what an **adapter** is. An adapter is a design pattern that manages the interface between two different modules, applications, languages, and so on. The title of the design pattern is descriptive of what we are doing. For instance, if you buy one of the new MacBook Pros, you will realize that you only have USB-C ports. Instead of opening your MacBook and rewiring it so that it can accept your standard USB memory stick, you buy an adapter. Adapters have multiple advantages. When it comes to modular software engineering, this gives us an advantage.

For instance, let's say that module A relies on module B. Instead of importing aspects of module B throughout module A, we can create adapters that manage the interface between both modules. This, in turn, gives us a lot of flexibility. For instance, module C could be built as an improvement on module B. Instead of working through module A looking for, and trying to root out, the uses of module B, we know that they are all utilized in the adapter. We can even produce a second adapter slowly moving over to module C in time. If we want to delete a module or move it out, again, our connection to another module can be severed instantly by just deleting the adapters. Adapters are simple and give us ultimate flexibility.

Considering what we have discussed about adapters, it makes sense that we create adapters between our Rust code and Python. Seeing as Python systems are essentially using our Rust code, it makes sense to build our adapters in Python.

To demonstrate how to do this, we will create an adapter that accepts either a list or an integer. It then selects the right Rust function and implements it. However, for our purposes with this adapter, we can make up a scenario where there is a lot of incorrect data being fed into the module. We do not want it to error every time incorrect data is passed in, but we do want to categorize whether the calculation is a failure, and we want to count the number of correct calculations we do. This seems specific, and we must remember that, like the MacBook, we can have multiple adapters. Nothing is stopping us from chopping, changing, and deleting in the future if we need to.

However, before we start writing code, we need to understand the layers involved for the adapter, as described here:

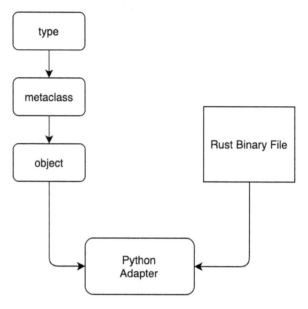

Figure 5.3 – The layers of a Python adapter for a Rust module

In the preceding figure, we can see that Python objects come from types. However, we can interject how these objects are called from types with **metaclasses**. When it comes to metaclasses, we must build a metaclass that will define how our counter is called. Our counter is going to be universal. We do not know how the users will use our interface. They might loop through a list of data points, calling our adapter for each one. We need to ensure that no matter how many adapters are being called, they are all pointing to the same counter. This might be a little confusing. This will become clearer when we build it.

Using a singleton design pattern to build an adapter interface

First, we must define our Singleton metaclass:

1. This can be done with the following code in the flitton_fib_rs/singleton.py file we create:

```
class Singleton(type):
    _instances = {}

    def __call__(cls, *args, **kwargs):
        if cls not in cls._instances:
```

```
        cls._instances[cls] = super(Singleton, \
            cls).__call__(*args, **kwargs)
    return cls._instances[cls]
```

2. Here, we can see that our `Singleton` class inherits directly from `type`. Here, what is happening is that we have a dictionary called `_instances`, where the keys to this dictionary are the class types. When a class that has `Singleton` as a metaclass is called, the type of that class is checked in the dictionary. If the type is not in the dictionary, then it is constructed and put into the dictionary. The instance in the dictionary is then returned. What this essentially means is that we cannot have two instances of a class. This process is laid out in the following figure:

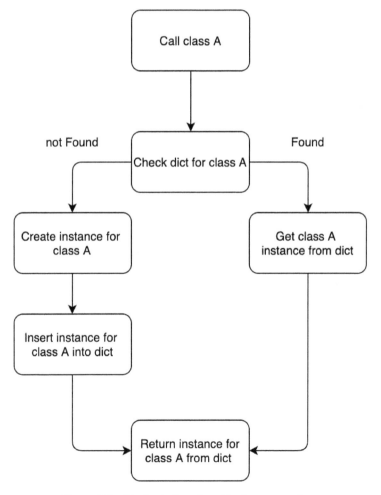

Figure 5.4 – The logic flow for a Singleton metaclass

3. Now, we will use our `Singleton` class to construct our counter. This can be done with the following code in the `flitton_fib_rs/counter.py` file that we create:

```python
from .singleton import Singleton


class Counter(metaclass=Singleton):

    def __init__(self, initial_value: int = 0) -> \
        None:
            self._value: int = initial_value

    def increase_count(self) -> None:
        self._value += 1

    @property
    def value(self) -> int:
        return self._value
```

Now, our `Counter` class cannot be constructed twice in the same program. Therefore, we can ensure that there will only be one `Counter` class, no matter how many times we call it.

4. We can now use it on our main adapter. We will house our main adapter in the `flitton_fib_rs/fib_number_adapter.py` file that we create. First of all, we import all of the functions and objects that we need with the following code:

```python
from typing import Union, List, Optional

from .flitton_fib_rs import fibonacci_number, \
    fibonacci_numbers
from .counter import Counter
```

Here, we can see that we have imported the typing that we need. We have also imported the Rust Fibonacci numbers that we will be using and our counter. Now that we have imported what we need, we can build our interface constructor.

5. For our adapter, we will need to have a number input, a status of whether the process is successful or not, along with the actual result, which will be the calculated Fibonacci number, or an error message if there is a failure. We will also have a counter, and we will have to process the input during the construction of the object. This can be denoted with the following code:

```
class FlittonFibNumberAdapter:

    def __init__(self,
        number_input: Union[int, List[int]]) -> None:
        self.input: Union[int, List[int]] = \
          number_input
        self.success: bool = False
        self.result: Optional[Union[int, List[int]]] \
          = None
        self.error_message: Optional[str] = None
        self._counter: Counter = Counter()
        self._process_input()
```

Remember, even though we call the counter, it is a singleton pattern; therefore, the counter will be the same instance across all instances of the adapter. Now that all of the correct attributes have been defined, we have to define what is an actual success.

6. This is where we state that success is `True`, and we increase the counter by one. This can be denoted by the `FlittonFibNumberAdapter` instance function, as shown here:

```
    def _define_success(self) -> None:
        self.success = True
        self._counter.increase_count()
```

This is smooth; because we have defined a clean interface for the counter, there is little explanation needed. Now that we have our success defined, we need to process the input because there are two different functions, one that takes a list and one that takes an integer.

7. We can pass in the correct input to the correct function with the
FlittonFibNumberAdapter instance function, as shown here:

```python
def _process_input(self) -> None:
    if isinstance(self.input, int):
        self.result = fibonacci_number( \
            n=self.input)
        self._define_success()

    elif isinstance(self.input, list):
        self.result = fibonacci_numbers( \
            numbers=self.input)
        self._define_success()
    else:
        self.error_message = "input needs to be \
        a list of ints or an int"
```

Here, we can see that we define an error message if there isn't a list of integers
passed in. If we do pass in the correct input, we define the result as the result of the
function and call the _define_success function.

8. The only thing left is to expose the count for the outside user. This can be done with
the following FlittonFibNumberAdapter property:

```python
@property
def count(self) -> int:
    return self._counter.value
```

9. Again, the counter interface is clean, so no explanation is needed. Our adapter
interface is now completed. All we need to do is expose it to the user by importing it
into the src/__init__.py file with the following code:

```python
from .fib_number_adapter import \
    FlittonFibNumberAdapter
```

Everything is done. We can now update our GitHub repository and reinstall our package
in the Python environment.

Testing our adapter interface in the Python console

We can now test our adapter with the Python console commands, as shown here:

```
>>> from flitton_fib_rs import FlittonFibNumberAdapter
>>> test = FlittonFibNumberAdapter(10)
>>> test_two = FlittonFibNumberAdapter(15)
>>> test_two.count
2
>>> test.count
2
>>> test_two.success
True
>>> test_two.result
610
```

Here, we can see that we can import our adapter from the module. We can then define two different adapters. We can see that the count is consistent across both adapters, which means that our singleton pattern works! Both adapters are pointing to the same Counter instance! All of our adapters will point to that same Counter instance. We can also see that the success is True, and we can access the result of the calculation:

1. Now, in the same Python console, we can test to see whether an incorrect input results in a failure and doesn't increase the count with the Python console commands shown next:

```
>>> test_three = FlittonFibNumberAdapter(
                        "should fail"
                 )
>>> test_three.count
2
>>> test_three.result
>>> test_three.success
False
>>> test_three.error_message
'input needs to be a list of ints or an int'
>>>
```

2. Here, we can see that the count hasn't increased, the success is `False`, and that there is an error message. The final input test can be done by inputting a list of integers with the Python console command shown next:

```
>>> test_four = FlittonFibNumberAdapter(
                        [5, 6, 7, 8, 9]
                )
>>> test_four.result
[5, 8, 13, 21, 34]
```

Here, we can see that it works. There is only one final test that we really know will work. However, it's to drive home what's happening with the singleton pattern. If we call all of the counts for all four adapters, they should all be 3, as they are all pointing to the same `Counter` instance and there was one failure out of the four.

3. Calling them in the same Python console command reveals whether this is true, as shown next:

```
>>> test.count
3
>>> test_two.count
3
>>> test_three.count
3
>>> test_four.count
3
```

There we have it. We have fully configured the Python interface of our module.

In this section, we built our Rust `pip` package with a Python interface. You might be tempted to add extra directories and flesh out entire Python modules in the `flitton_fib_rs` directory. However, extra directories in the `flitton_fib_rs` directory do not get copied over when the package is being installed. This is fine as well. We are essentially building Rust `pip` packages. Rust is fast and safe, and we should be leaning on this as much as we can. The Python adapters and command in the `flitton_fib_rs` directory should be there to smooth over the interface. For instance, if we want the memory of our interface to be managed in a particular way, it makes sense to do this in the interface of Python as a wrapper, as Python will be the system that is importing and using the `pip` package. If you find yourself putting anything other than adapters and command-line functions in the `flitton_fib_rs` module, that is a warning sign that you should try and consider putting it in the Rust module itself. We have tested our package manually; however, we need to ensure that our Rust Fibonacci calculation functions do as we expect.

In the next section, we will be creating unit tests for our Rust code.

Building tests for our Rust package

Previously, in *Chapter 4*, *Building pip Modules in Python*, we built unit tests for our Python code. In this section, we will build unit tests for our Fibonacci functions. These tests do not need any extra packages or dependencies. We can use Cargo to manage our testing. This can be done by adding our testing code in the `src/fib_calcs/fib_number.rs` file. The steps are as follows:

1. We do this by creating a module in the `src/fib_calcs/fib_number.rs` file with the following code:

    ```
    #[cfg(test)]
    mod fibonacci_number_tests {
        use super::fibonacci_number;
    }
    ```

    Here, we can see that we have defined a module in the same file and decorated the module with the `#[cfg(test)]` macro.

2. We can also see that we must import the function, as it is super to the module. Inside this module, we can run standard tests that check to see whether the integers we pass in calculate the Fibonacci number we expect with the following code:

    ```
    #[test]
    fn test_one() {
        assert_eq!(fibonacci_number(1), 1);
    }
    #[test]
    fn test_two() {
        assert_eq!(fibonacci_number(2), 1);
    }
    #[test]
    fn test_three() {
        assert_eq!(fibonacci_number(3), 2);
    }
    #[test]
    fn test_twenty() {
    ```

```
        assert_eq!(fibonacci_number(20), 6765);
    }
```

Here, we can see that we have decorated our test functions with the #[test] macro. If they do not produce the results that we expect, then assert_eq! and the test will fail. We also must note that our function will panic if we pass in zero or a negative value.

3. These can be tested with the test functions, as shown next:

```
#[test]
#[should_panic]
fn test_0() {
    fibonacci_number(0);
}
#[test]
#[should_panic]
fn test_negative() {
    fibonacci_number(-20);
}
```

Here, we pass in the failing inputs. If they do not panic, then the test will fail because we decorated it with the #[should_panic] macro.

4. Now that we have created our tests for the fibonacci_number function, we can build our test for our fibonacci_numbers function in the src/fib_calcs/ fib_numbers.rs file with the following code:

```
#[cfg(test)]
mod fibonacci_numbers_tests {

    use super::fibonacci_numbers;

    #[test]
    fn test_run() {
    let outcome = fibonacci_numbers([1, 2, 3, \
        4].to_vec());
        assert_eq!(outcome, [1, 1, 2, 3]);
    }
}
```

5. Here, we can see that this has the same layout as our other tests. If we want to run our tests, we can run them with the following command:

```
cargo test
```

This gives us the following printout:

```
running 7 tests
test fib_calcs::fib_number::fibonacci_number_tests::test_
th

ree ... ok
test fib_calcs::fib_numbers::fibonacci_numbers_
tests::test_

run ... ok
test fib_calcs::fib_number::fibonacci_number_tests::

test_two ... ok
test fib_calcs::fib_number::fibonacci_number_tests::test_
on

e ... ok
test fib_calcs::fib_number::fibonacci_number_tests::

test_twenty ... ok
test fib_calcs::fib_number::fibonacci_number_tests::

test_negative ... ok
test fib_calcs::fib_number::fibonacci_number_tests::

test_0 ... ok
test result: ok. 7 passed; 0 failed; 0 ignored; 0

measured; 0
filtered out; finished in 0.00s
        Running target/debug/deps/flitton_fib_rs-
07e3ba4b0bc8cc1e
running 0 tests
test result: ok. 0 passed; 0 failed; 0 ignored; 0

measured; 0
filtered out; finished in 0.00s
    Doc-tests flitton_fib_rs
running 0 tests
test result: ok. 0 passed; 0 failed; 0 ignored; 0

measured; 0
filtered out; finished in 0.00s
```

Here, we can see that all of our tests have run and passed. If we recall *Chapter 4, Building pip Modules in Python*, we'll remember that we used mocking.

Rust is still developing **mock crates**. One crate, `mockall`, enables mocking and can be found at this URL: `https://docs.rs/mockall/0.10.0/mockall/`. Another cleaner crate that can be utilized for mocking can be found at this URL: `https://docs.rs/mocktopus/0.7.11/mocktopus/`.

We have now covered how to build our module and test it. We are at the end of building a Rust `pip` module with tests and a Python interface. We can now test the speed of our Rust module to see what will happen and how powerful Rust modules as a tool are.

Comparing speed with Python, Rust, and Numba

We have now built a `pip` module in Rust with command-line tools, Python interfaces, and unit tests. This is a shiny new tool that we have. Let's put it to the test. We know that Rust by itself is faster than Python. However, do we know that the `pyo3` bindings slow us down? Also, there is another way to speed up our Python code and this is with Numba, a Python package that compiles Python code to speed it up. Should we go through all of the haste of creating the Rust package if we can achieve the same speed with Numba? In this section, we will run our Fibonacci function several times, in Python, Numba, and our Rust module. It has to be noted that Numba can be a headache to install. For instance, I could not install it on my MacBook Pro M1. I had to install Numba on a Linux laptop to run this section. You don't have to run the code in this section; it is more for demonstrative purposes. If you do want to try and run the test script, then all of the steps are provided:

1. First of all, we have to install the Rust `pip` module that we have built. We then install Numba with the following command:

    ```
    pip install numba
    ```

2. Once this is done, we have everything we need. In any Python script, we import the packages required with the following code:

    ```
    from time import time

    from flitton_fib_rs.flitton_fib_rs import \
        fibonacci_number
    from numba import jit
    ```

We are using the `time` module to time how long it takes for each run to happen. We also use the Fibonacci function from our Rust `pip` module, and we also require the `jit` decorator from Numba. **jit** stands for **just in time**. This is because Numba compiles the function when it loads it.

3. We now define our standard Python function with the following code:

```
def python_fib_number(number: int) -> int:
    if number < 0:
        raise ValueError(
            "Fibonacci has to be equal or above zero"
        )
    elif number in [1, 2]:
        return 1
    else:
        return numba_fib_number(number - 1) + \
            numba_fib_number(number - 2)
```

4. We can see that this is the same logic that the Rust code is built with. We want to ensure that our tests are reputable comparisons. We then define the Python function that is compiled with `jit` with the following code:

```
@jit(nopython=True)
def numba_fib_number(number: int) -> int:
    if number < 0:
        raise ValueError("Fibonacci has to be equal \
            or above zero")
    elif number in [1, 2]:
        return 1
    else:
        return numba_fib_number(number - 1) + \
            numba_fib_number(number - 2)
```

5. We can see that it is the same. The only difference is that we have decorated it with `jit` and set `nopython` to `True` to obtain optimal performance. We then run all of them with the following code:

```
t0 = time()
for i in range(0, 30):
    numba_fib_number(35)
```

```
t1 = time()
print(f"the time taken for numba is: {t1-t0}")
t0 = time()
for i in range(0, 30):
    numba_fib_number(35)
t1 = time()
print(f"the time taken for numba is: {t1 - t0}")
```

Here, we can see that we loop through a range from 0 to 30 and hit our function 30 times with the number 35. We then print the time elapsed for this to happen. We notice that we have done this twice. This is because the first run will involve compiling the function.

6. When we run this, we get the following console printout:

```
the time taken for numba is: 2.6187334060668945
the time taken for numba is: 2.4959869384765625
```

7. Here, we can see that some time is shaved off in the second run because it is not compiling. Running this several times shows that this reduction is standard. Now, we set up our standard Python test with the following code:

```
t0 = time()
for i in range(0, 30):
    python_fib_number(35)
t1 = time()
print(f"the time taken for python is: {t1 - t0}")
```

8. Running this test will get the following console printout:

```
the time taken for python is: 2.889884853363037
```

9. We can see that there is a significant speed decrease when it comes to running pure Python code as opposed to our Numba function. Now, we can move on to the final test, which is our Rust test, defined with the following code:

```
t0 = time()
for i in range(0, 30):
    fibonacci_number(35)
t1 = time()
print(f"the time taken for rust is: {t1 - t0}")
```

10. Running this test gives us the following console printout:

```
the time taken for rust is: 0.9373788833618164
```

Here, we can see that the Rust function is a lot faster. This does not mean that Numba is a waste of time. When it comes to Python optimizations, Numba can perform well in certain situations. In other situations, the Python optimizations will not affect them at all. Considering how easy they are to apply, it is always worth checking to see whether there is a speed-up. However, we also now know that Rust will always be faster than pure Python code.

Summary

In this chapter, we have built a fully fleshed-out Python `pip` module with command-line tools, interfaces, and Rust code. We managed `gitignore` for both Rust and Python development. We then defined our setup tools for packaging our Python code and module with the compilation of Rust code that has Python bindings. Once these were defined, we learned how to build Rust functions that spanned multiple Rust files that could be wrapped in `pyo3` bindings.

Our development did not just stop at Rust. We also explored Python's singleton and adapter design patterns to build more advanced Python interfaces for our users. We then tested our code with unit tests and speed checking. It must be noted that we did not cover GitHub actions in this chapter. GitHub actions are defined in the same way as they were in the previous chapter. Instead of running tests using the Python unit test, we run our tests using Cargo and so on. However, uploading to PyPI is a little more complicated. To cover this, examples on how to pre-compile and upload Rust `pip` modules are provided in the *Further reading* section.

We now have a powerful skill, which is building Python `pip` modules that utilize Rust. However, we leaned on our Python to build our interfaces. In the next chapter, we will work with Python objects within our Rust code. Therefore, we will be able to pass in more advanced Python data objects into our Rust code. We will also enable our Rust code to return fully fledged Python objects.

Questions

1. How do you define a `setup.py` file for a `pyo3` Rust Python `pip` module?

2. What is the layout of our `pip` module in the Python environment after it has been installed? Also, why can we not build Python modules spanning multiple directories?

3. What is a singleton design pattern?

4. What is an adapter design pattern and what are the advantages of using the design pattern?

5. What is a metaclass and how do we use it?

Answers

1. Here, we must use the `dist` package to install `setuptools_rust` before we do anything else in the `setup.py` file. We define the parameters for the setup and use the `RustExtension` object from `setuptools_rust`, pointing to where the compiled Rust module will be once installed.

2. When the `pip` module is installed, the binary Rust file is in the same directory where the Python files are defined for the module. However, directories in that directory are not copied over and, therefore, they will be lost during the installation.

3. A singleton design pattern ensures that all references to a particular class all point to one instance of that class.

4. The adapter pattern is an interface that manages the interaction between two modules. The advantage is the flexibility between the modules. We know where all the interactions are, and if we want to sever the modules, all we need to do is delete the adapter. This enables us to switch modules as and when we need them.

5. A metaclass is a class that lies between a type and an object. Because of this, we can use this to see how we manage calling our objects.

Further reading

- *Mre – an example of GitHub actions for deploying Rust packages on PyPI (2021)*: `https://github.com/mre/hyperjson/blob/master/.github/workflows/ci.yml`

- *Mastering Object-Oriented Python, Steven F. Lott, Packt Publishing* (2019)

- *The PyO3 user guide*: `https://pyo3.rs/v0.13.2/`

6

Working with Python Objects in Rust

So far, we have managed to fuse Rust with Python to speed up our code. However, software programs written in Rust can get complicated. While we can get by with passing integers and strings into Rust functions from Python code, it would be useful to handle more complex data structures from Python and objects. In this chapter, we accept and process Python data structures such as a **dictionary**. We will go further by processing custom Python objects and even creating Python objects inside our Rust code.

In this chapter, we will cover the following topics:

- Passing complex Python objects into Rust
- Inspecting and working with custom Python objects
- Constructing our own custom Python objects in Rust

Technical requirements

The code for this chapter can be found via the following GitHub link:

`https://github.com/PacktPublishing/Speed-up-your-Python-with-Rust/tree/main/chapter_six`

Passing complex Python objects into Rust

A key skill that enables us to take our Rust `pip` module development to the next level is taking in complex Python data structures/objects and using them. In *Chapter 5, Creating a Rust Interface for Our pip Module*, we accepted integers. We noticed that these raw integers were just directly transferred to our Rust function. However, with Python objects, it is more complex than this.

To explore this, we will create a new command-line function that reads a .yml file and passes a Python dictionary into our Rust function. The data in this dictionary will have the parameters needed for firing our `fibonacci_numbers` and `fibonacci_number` Rust functions, adding the results of those functions to the Python dictionary and passing it back to the Python system.

To achieve this, we must carry out the following steps:

1. Update our `setup.py` file to support .yml loading and a command-line function that reads it.
2. Define a command-line function that reads the .yml file and feeds it into Rust.
3. Process data from our Python dictionary for `fibonacci_numbers` in Rust.
4. Extract data from our config file.
5. Return our Python dictionary to our Python system.

This approach will require us to write the whole process before we can run it. This can be frustrating because we cannot see it working until the end. However, it is laid out in this book this way so that we can see the data flow. We are exploring the concept of passing complex data structures into Rust for the first time. Once we understand how this works, we can then develop `pip` modules that work for us as individuals.

Updating our setup.py file to support .yml loading

Let's start this journey by updating our `setup.py` file, as follows:

1. With our new command-line function, we read a .yml file and pass through that data to our Rust function. This requires our Python `pip` module to have the `pyyaml` Python module. This can be done by adding the `requirements` parameter to our `setup` initialization, as follows:

```
requirements=[
    "pyyaml>=3.13"
]
```

We remember that we can keep adding more dependencies to our module by just adding them to our `requirements` list. If we want our module to be more flexible for multiple installs to different systems, it is advised that we can lower the version number for our `pyyaml` module requirement.

2. Now that we have defined our requirements, we can define a new console script, resulting in the `entry_points` parameter in our `setup` initialization, which looks like this:

```
entry_points={
    'console_scripts': [
        'fib-number = flitton_fib_rs.'
        'fib_number_command:'
        'fib_number_command',
        'config-fib = flitton_fib_rs.'
        'config_number_command:'
        'config_number_command',
    ],
},
```

With this, we can see that our new console script will be in the `flitton_fib_rs/config_number_command.py` directory.

3. In the `flitton_fib_rs/config_number_command.py` directory, we need to build a function called `config_number_command`. First, we need to import the required modules, as follows:

```
import argparse
import yaml
import os
```

```
from pprint import pprint

from .flitton_fib_rs import run_config
```

os will help us with the path definition to the .yml file. The pprint function will just help us print the data in an easy-to-read format on the console. We have also defined a Rust function that will process our dictionary as run_config.

Defining our .yml loading command

Now that our imports have been done, we can define our function and collect the command-line arguments. Here's how we do this:

1. You can start with the following code:

```
def config_number_command() -> None:
    parser = argparse.ArgumentParser(
        description='Calculate Fibonacci numbers '
                    'using a config file')
    parser.add_argument('--path', action='store',
                            type=str, required=True,
                            help="path to config file")
    args = parser.parse_args()
```

2. Here, we can see that we take in a string, which is the path to the .yml file with the --path tag, and we parse it. Now that we have parsed the path, we can open our .yml file by running the following code:

```
with open(str(os.getcwd()) + "/" + args.path) as \
    f:
        config_data: dict = yaml.safe_load(f)
```

Here, we can see that we attach our path with the os.getcwd() function. This is because we must know where the user is calling the command. For instance, if we are in the x/y/ directory and we want to point to the x/y/z.yml file, we will have to run the config-fib --path z.yml command. If the directory of the file were x/y/test/z.yml, we would have had to run the config-fib --path test/z.yml command.

3. Now that we have our data loaded from the `.yml` file, we can print it out and print out the results of our Rust function by running the following code:

```
print("Here is the config data: ")
pprint(config_data)
print(f"Here is the result:")
pprint(run_config(config_data))
```

With this, we have now completed all our Python code.

Processing data from our Python dictionary

We are now going to have to build Rust functions that process the Python dictionaries. Here's how we'll go about this:

1. When it comes to processing input dictionaries, we must agree on a format that we are going to accept. To keep this simple, our Python dictionaries will have two keys. The `number` key is for a list of integers that can call Fibonacci number calculations individually, while the `numbers` key is for a list of lists of integers. To ensure that our Rust code does not become disorganized, we are going to define our interfaces in our own interface directory, giving our Rust code the following structure:

```
├── fib_calcs
│   ├── fib_number.rs
│   ├── fib_numbers.rs
│   └── mod.rs
├── interface
│   ├── config.rs
│   └── mod.rs
├── lib.rs
└── main.rs
```

2. We will build our configuration interface in the `src/interface/config.rs` file. First, we are going to import all the functions and macros that we need, as follows:

```
use pyo3::prelude::{pyfunction, PyResult};
use pyo3::types::{PyDict, PyList};
use pyo3::exceptions::PyTypeError;
```

```
use crate::fib_calcs::fib_number::fibonacci_number;
use crate::fib_calcs::fib_numbers::fibonacci_numbers;
```

We are going to use pyfunction to wrap our interface that takes in a Python dictionary. We will return the dictionary back to the Python program wrapped in a pyResult struct. Seeing as we are accepting a Python dictionary, we will be using a PyDict struct to describe the dictionary being passed in and returned. We will also be accessing the lists in the dictionary using a PyList struct. If there is an issue with our dictionary not housing lists, then we will have to throw an error that the Python system will understand. To do this, we will use a PyTypeError struct. Finally, we will be using our Fibonacci number functions to calculate the Fibonacci numbers. We can see that we are simply importing from another module in the Rust code with use crate::. Even though our Fibonacci number functions have the pyfunction macro applied to them, nothing is stopping us from using them as normal Rust functions elsewhere in our Rust code.

3. Before we write our interface function, we need to build a private function that accepts our lists of lists, calculates the Fibonacci numbers, and returns them in a list of lists, as seen in the following code snippet:

```
fn process_numbers(input_numbers: Vec<Vec<i32>>) \
    -> Vec<Vec<u64>> {
    let mut buffer: Vec<Vec<u64>> = Vec::new();
    for i in input_numbers {
        buffer.push(fibonacci_numbers(i));
    }
    return buffer
}
```

4. This should be straightforward at this stage in the book. Considering this, we now have everything we need to build our interface. First, we need to define a pyfunction function that accepts and returns the same data by running the following code:

```
#[pyfunction]
pub fn run_config<'a>(config: &'a PyDict) \
    -> PyResult<&'a PyDict> {
```

Here, we can see that we tell the Rust compiler that the Python dictionary that we accept must have the same lifetime as the Python dictionary that we are returning. This makes sense as we are returning the same dictionary after adding the results to it.

5. Our first process is to see if the number key is present in the dictionary by running the following code:

```
match config.get_item("number") {
    Some(data) => {

        . . .

    },
    None => println!(
    "parameter number is not in the config"
    )
}
```

Here, we can see that is the number key is not there, so we merely print that it is not there. We can change the rules to throw an error instead, but we are accepting a forgiving config file. If the user does not have any individual Fibonacci numbers to compute, only lists of them, then we should not throw errors, insisting that the user adds the field. The three dots in the code snippet shown in *Step 6* are where the code is going to be executed if the number key is present.

6. We substitute the three dots in the following code snippet:

```
match data.downcast::<PyList>() {
    Ok(raw_data) => {

        . . .

    },
    Err(_) => Err(PyTypeError::new_err(
        "parameter number is not a list
        of integers")).unwrap()
}
```

Here, we can see that we downcast the data we extracted belonging to the `number` key to the `PyList` struct. If this fails, then we actively throw a type error because the user has tried to configure the `number` key but failed. If it passes, we can run the Fibonacci function by substituting the three dots in the preceding code snippet with the following code:

```
let processed_results: Vec<i32> =
    raw_data.extract::<Vec<i32>>().unwrap();
let fib_numbers: Vec<u64> =
    processed_results.iter().map(
        |x| fibonacci_number(*x)
    ).collect();
config.set_item(
    "NUMBER RESULT", fib_numbers);
```

Here, what we have done is create `Vec<i32>` by running the `extract` function on the `PyList` struct. We directly unwrap it so that if there is an error, it will be thrown straight away. We then create `Vec<u64>`, which houses the calculated Fibonacci numbers, by iterating through the vector with the `iter()` function. We then map each `i32` integer of that vector with the `map` function. Inside the `map` function, we define a closure that is mapped to each `i32` integer in the vector. It must be noted that we apply the `fibonacci` function where we dereference the `i32` integer being passed in because it is now a borrowed reference. We collect the results of this mapping with the `.collect()` function, which results in the `processed_results` variable being a collection of `i32` calculated Fibonacci numbers. We then add the calculated numbers to the dictionary under the NUMBER RESULT key. We can see the flow of what was just described in the following diagram:

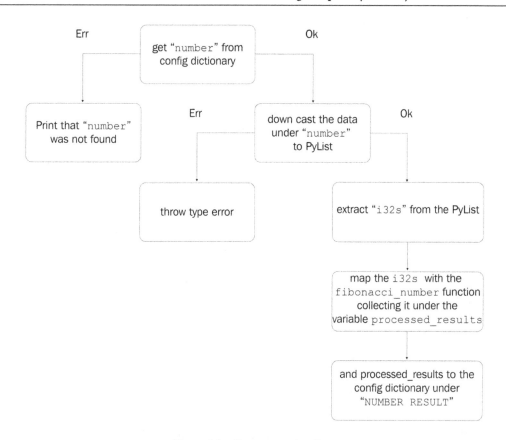

Figure 6.1 – Data extraction flow

In the next step, we will carry out a similar process to the one displayed in *Figure 6.1* to process the list of lists under the numbers key.

Extracting data from our config file

At this point, it would be a good idea to try to implement the process for the numbers key by yourself. To make things easier, you can use the process_numbers function that we defined earlier in *Step 3* of the *Processing data from our Python dictionary* section. We will cover the solution to this in the next steps:

1. The numbers key can be processed by our run_config function with the code defined here:

```
match config.get_item("numbers") {
    Some(data) => {
        match data.downcast::<PyList>() {
```

```
            Ok(raw_data) => {
                let processed_results_two: \
                    Vec<Vec<i32>> =
                raw_data.extract::<Vec<Vec<i32>>>(
                ).unwrap();
                config.set_item("NUMBERS RESULT",
                process_numbers(processed \
                    _results_two));
            },
            Err(_) => Err(PyTypeError::new_err(
            "parameter numbers is not a list of \
                lists of integers")).unwrap()
        }

        },
    None => println!(
        "parameter numbers is not in the config")
    }
    return Ok(config)
```

Here, we can see that the `process_numbers` function actually makes this implementation simpler than the `numbers` key processing. If the complexity starts to grow, it is always worth breaking down the logic into smaller functions. It also must be noted that we return a result that wraps the config dictionary. Now that we have finished the logic behind processing our dictionary, we need to return our dictionary in the next step.

2. Here, we must publicly define our `src/interface/config.rs` file in the `src/interface/mod.rs` file by running the following code:

```
pub mod config;
```

3. We then import it into our `src/lib.rs` file by running the following code:

```
mod interface;
```

```
use interface::config::__pyo3_get_function_run_config;
```

4. We then add the function to our module in the `src/lib.rs` file by running the following code:

```
m.add_wrapped(wrap_pyfunction!(run_config));
```

We have now carried out all the steps.

Returning our Rust dictionary to our Python system

Our `pip` module can now take in a configuration file, convert it into a Python dictionary, pass the Python dictionary into the Rust function that calculates the Fibonacci numbers, and return the results in the form of a dictionary back to Python. This can be achieved by carrying out the following steps:

1. Define a `.yml` file to be ingested by our program. An example `.yml` file that can run what we have just done can be defined via the following code:

```
number:
  - 4
  - 7
  - 2
numbers:
  -
    - 12
    - 15
    - 20
  -
    - 15
    - 19
    - 18
```

I have saved the preceding .yml code on my desktop for demonstration purposes under the filename `example.yml`. Remember to update your GitHub repository and uninstall your current module in your Python environment, and install our new module instead.

2. We can then pass in the `.yml` file into our module entry point with the following command:

```
config-fib --path example.yml
```

3. I ran this command from my desktop, where I stored the `example.yml` file. Running the previous command gives us the following output:

```
Here is the config data:
{'number': [4, 7, 2, 10, 15],
 'numbers': [[5, 8, 12, 15, 20], [12, 15, 19, 18, 8]]}
Here is the result:
{'NUMBER RESULT': [3, 13, 1, 55, 610],
 'NUMBERS RESULT': [[5, 21, 144, 610, 6765],
                    [144, 610, 4181, 2584, 21]],
 'number': [4, 7, 2, 10, 15],
 'numbers': [[5, 8, 12, 15, 20], [12, 15, 19, 18, 8]]}
```

Here, we can see that our Python interface fed the Python dictionary into the Rust interface. We then got the results of the Fibonacci functions passed back in the same dictionary.

4. Now, we introduce a breaking change in our `.yml` file. We can test our error by changing the `number` key to a dictionary as opposed to a list of integers in our `example.yml` file by running the following code:

```
number:
    one: 1
```

5. Finally we run our code again, expecting the correct error message. This gives us the following error when running our command again:

```
pyo3_runtime.PanicException: called 'Result::unwrap()'
on an 'Err' value: PyErr { type: <class 'TypeError'>,
value: TypeError('parameter number is not a list of
integers'),
traceback: None }
```

Here, we can see that a `TypeError` exception was raised. This is not trivial. This means that we can try to accept type errors in our Python code when using our Rust module if we need to. Considering this, if a user did not know how our module was built, they would have no problem thinking that our module was built in pure Python. There is one more test that we can consider. We only manually threw an error when we were downcasting to `PyList`, highlighting that we need to have a list of integers. However, we just unwrapped the `extract` function being performed on `PyList`.

6. We can see how the `extract` function handles a string being put in, thereby changing the `number` key to a list of strings as opposed to a list of integers in our `example.yml` file, by running the following code:

```
number:
  - "test"
```

7. Running our command again gives us the following output:

```
pyo3_runtime.PanicException: called 'Result:: \
  unwrap()' on an
'Err' value: PyErr { type: <class 'TypeError'>,
value: TypeError(
"'str' object cannot be interpreted as an integer"),
traceback: None }
```

Here, we can see that the error string is a little harder to interpret because we did not directly code an error telling the user what we want; however, it is still `TypeError`. We can also see here that errors raised by functions that are acted on Python objects are Python-friendly.

We have now concluded how to interact with complex Python data structures. Nothing is stopping you from building Python `pip` modules in Rust that fuse seamlessly with a Python program. However, we can take our Rust `pip` modules to the next level by working with and inspecting custom Python objects in the next section.

Inspecting and working with custom Python objects

Technically, everything in Python is an object. The Python dictionary that we worked on within the previous section is an object, so we have already managed Python objects. However, as we know, Python enables us to build custom objects. In this section, we will get our Rust function to accept a custom Python class that will have `number` and `numbers` attributes. To achieve this, we must carry out the following steps:

1. Create an object that passes itself into our Rust interface.
2. Acquire the Python **global interpreter lock (GIL)** within our Rust code to create a `PyDict` struct.
3. Add the custom object's attributes to our newly created `PyDict` struct.
4. Set the attributes of the custom object to the results of our `run_config` function.

Creating an object for our Rust interface

We start our journey by setting up our interface object, as follows:

1. We house our object that will pass itself into our Rust code in the `flitton_fib_ rs/object_interface.py` file. Initially, we import what we need by running the following code:

    ```python
    from typing import List, Optional


    from .flitton_fib_rs import object_interface
    ```

2. We then define the `__init__` method of our object by running the following code:

    ```python
    class ObjectInterface:

        def __init__(self, number: List[int], \
            numbers: List[List[int]]) -> None:
            self.number: List[int] = number
            self.numbers: List[List[int]] = numbers
            self.number_results: Optional[List[int]] = \
                None
            self.numbers_results:Optional[List[List \
                [int]]] = None
    ```

    Here, we can see that we can pass in the Fibonacci numbers that we want to be calculated in the parameters. We then just set our attributes to the parameters that we passed in. The result parameters defined here are of a None value. However, they will be populated by the Rust code when we pass this object into our Rust object interface.

3. We then define a function that will pass our object into the Rust code by running the following code:

    ```python
    def process(self) -> None:
        object_interface(self)
    ```

Here, we can see that this is done by merely passing the `self` reference into the function. Now that we have defined our object, we can move on to build our interface and interact with the Python GIL.

Acquiring the Python GIL in Rust

For our interface, we will house our function in the `src/interface/object.rs` file. We'll proceed as follows:

1. First, we must import all of what we need by running the following code:

```
use pyo3::prelude::{pyfunction, PyResult, Python};
use pyo3::types::{PyAny, PyDict};
use pyo3::exceptions::PyLookupError;


use super::config::run_config;
```

    Most of these imports will be familiar by now. The new import that we must make note of is the `Python` import. `Python` is a struct that is essentially a marker that is required for the Python operations that we will be doing.

2. Now that we have imported everything we need, we can build parameters for our interface and create a `PyDict` struct by running the following code:

```
#[pyfunction]
pub fn object_interface<'a>(input_object: &'a PyAny) \
    -> PyResult<&'a PyAny> {
    let gil = Python::acquire_gil();
    let py = gil.python();

    let config_dict: &PyDict = PyDict::new(py);
```

Here, what we have essentially done is acquire the Python GIL, and then use this to create a `PyDict` struct. To fully understand what we are doing, it is best to explore what the Python GIL is. In *Chapter 3, Understanding Concurrency*, we covered the concept of thread blocking. This means that if another thread is executing, then all other threads are locked. The GIL ensures that this happens, as demonstrated in the following diagram:

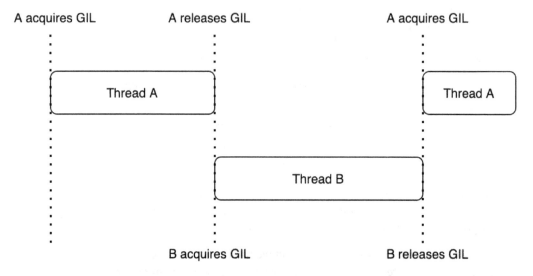

Figure 6.2 – GIL flow

This is because Python does not have any concept of ownership. A Python object can be referenced as many times as we want. We can also mutate the variable from any of those references. When we acquire the `gil` variable, we ensure that only one thread can use the Python interpreter and the Python **application programming interface** (**API**) at the same time. We must remember that processes behave differently and have their own memory. Our `gil` variable is a `GILGuard` struct that ensures that we acquire the GIL before we run any operations on Python objects.

Adding data to our newly created PyDict struct

Now that we have control over Python objects with the GIL, we can move on to our next step, where we add the data from the input object to our newly created PyDict struct, as follows:

1. Our approach in this step can be summarized in the following diagram:

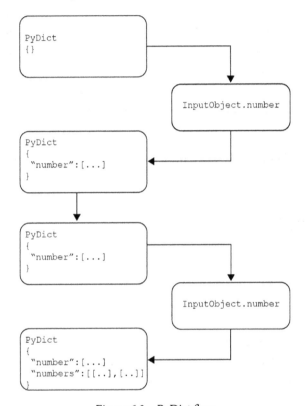

Figure 6.3 – PyDict flow

2. We can achieve the first cycle depicted in *Figure 6.3* by running the following code:

```
match input_object.getattr("number") {
    Ok(data) => {
        config_dict.set_item("number", data) \
            .unwrap();
    },
    Err(_) => Err(PyLookupError::new_err(
        "attribute number is missing")).unwrap()
}
```

Here, we can see that we match the `getattr` function, throwing an error if `input_object` does not have the `number` attribute. If we do have the attribute, we assign it to `config_dict`.

3. We can do the second cycle by running the following code:

```
match input_object.getattr("numbers") {
    Ok(data) => {
        config_dict.set_item("numbers", data) \
            .unwrap();
    }
    Err(_) => Err(PyLookupError::new_err(
        "attribute numbers is missing")).unwrap()
}
```

4. It must be noted that there is a fair amount of repetition, with only one change. We could refactor this into a single function with an `attribute` parameter by running the following code:

```
fn extract_data<'a>(input_object: &'a PyAny, \
    attribute: &'a str, config_dict: &'a PyDict) \
        -> &'a PyDict {
    match input_object.getattr(attribute) {
        Ok(data) => {
            config_dict.set_item(attribute, \
                data).unwrap();
        },
        Err(_) => Err(PyLookupError::new_err(
            "attribute number is missing")).unwrap()
    }
    return config_dict
}
```

5. Here, we can see that we get a lot of flexibility with our Python objects. This function can be used multiple times with the refactored code in our `object_interface` function, as seen here:

```
let mut config_dict: &PyDict = PyDict::new(py);
config_dict = extract_data(input_object, \
    "number", config_dict);
```

```
config_dict = extract_data(input_object,
   "numbers", config_dict);
```

Here, we can see that we have changed `config_dict` to a mutable. Now that we have loaded our `PyDict` struct with all the data that we need, all we must do is run our `run_config` function, add it to the input object's attributes, and return it to the Python interface in the next step.

Setting the attributes of our custom object

We are now in the final stage of our interface module. Here are the steps:

1. We can pass the output from our `run_config` function to our Python object interface by running the following code:

    ```
    let output_dict: &PyDict = run_config( \
       config_dict).unwrap();

    input_object.setattr(
        "number_results",
        output_dict.get_item(
            "NUMBER RESULT").unwrap()).unwrap();

    input_object.setattr(
        "numbers_results",
        output_dict.get_item(
            "NUMBERS RESULT").unwrap()).unwrap();

    return Ok(input_object)
    ```

    Here, we can see that we get the `output_dict` Python dictionary from the `run_config` function. Once we have got this, we set the `input_object` attribute based on the items from `output_dict`.

2. We have now completed our interface and we must subsequently plug it into our Rust module. We publicly define our interface file in the `src/interface/mod.rs` file by running the following code:

    ```
    pub mod object;
    ```

3. We can then define our interface function in our Rust module by importing it into our `src/lib.rs` file, as follows:

```
use interface::object::__pyo3_get_function_object_ \
    interface;
```

4. We then add our function to our module, as follows:

```
m.add_wrapped(wrap_pyfunction!(object_interface));
```

5. Our module is now fully functioning. As always, we must remember to update our GitHub repository, uninstall our old module in our Python environment, and reinstall it. Once this is done, we can test it by running a Python shell. In our shell, we can test our object by running the following code:

```
>>> from flitton_fib_rs.object_interface import
ObjectInterface
>>> test = ObjectInterface([5, 6, 7, 8], [])
>>> test.process()
>>> test.number_results
[5, 8, 13, 21]
```

Here, we can see that we import the object that we are going to use. We then initialize it and run the `process` function. Once this is done, we can see that our Rust code accepted our object and interacted with it as we have the correct results for our `number_results` attribute.

Now we can interact with Python custom objects, the problems we can solve and how we can interact with the Python system are powerful. Custom Python objects do not hold us back. However, it is important not to get too carried away with Python objects in our Rust code. While we should use them in our interface, we shouldn't have to rely on them to build the whole program. In this section, we did do this, as we were leaning on a function that we built in the previous section to avoid excessive code, to get a point across. However, in your projects, Python objects should leave your code after the interface. If you find yourself using Python objects in your Rust code throughout, then you must ask yourself why you are not just using pure Python. Coding in Python will be slower than coding in Rust, but the metaclass, dynamic attributes, and many other Python features will make coding in Python easier and more enjoyable than trying to force a Python style of coding into Rust. Rust offers structs, traits, enums, and strong typing with lifetimes that get cut after moving out of scope to keep resources low.

So, lean into this style of coding to fully reap the benefits of building `pip` modules in Rust. Push past your comfort zone of the Python coding style. The next section is about building Python objects in Rust code.

Constructing our own custom Python objects in Rust

In this final section, we will build a Python module in Rust that can be interacted with in the Python system as if it were a native Python object. To do this, we must follow these steps:

1. Define a Python class with all our attributes.

2. Define class static methods to process numbers.

3. Define a class constructor.

Defining a Python class with the required attributes

To start our journey, we define our class in the `src/class_module/fib_processor.rs` file, as follows:

1. To build our class, we need to import the required macros by running the following code:

   ```
   use pyo3::prelude::{pyclass, pymethods, staticmethod};

   use crate::fib_calcs::fib_number::fibonacci_number;
   use crate::fib_calcs::fib_numbers::fibonacci_numbers;
   ```

   Here, we are using the `pyclass` macro to define our Rust Python class. We then use `pymethods` and `staticmethod` to define methods attached to the class. We also use standard Fibonacci numbers to calculate the Fibonacci numbers.

2. Now that we have imported everything we need, we can define the class and the attributes, as follows:

   ```
   #[pyclass]
   pub struct FibProcessor {
       #[pyo3(get, set)]
       pub number: Vec<i32>,
       #[pyo3(get, set)]
       pub numbers: Vec<Vec<i32>>,
   ```

```
    #[pyo3(get)]
    pub number_results: Vec<u64>,
    #[pyo3(get)]
    pub numbers_results: Vec<Vec<u64>>
}
```

Here, we can see that we use Rust typing for our attributes. We also use a macro to state what we can do with these attributes. For our `number` and `numbers` attributes, we can get and set data belonging to these attributes. However, with our `results` attributes, we can only get data as this is set by the calculations.

Defining class static methods to process input numbers

We can now use our attributes to implement class methods.

Just as with standard structs, we can implement methods attached to the class with an `impl` block, as seen in the following code snippet:

```
#[pymethods]
impl FibProcessor {

    #[staticmethod]
    fn process_numbers(input_numbers: Vec<Vec<i32>>) \
        -> Vec<Vec<u64>> {
            let mut buffer: Vec<Vec<u64>> = Vec::new();
            for i in input_numbers {
                buffer.push(fibonacci_numbers(i));
            }
            return buffer
        }
}
```

Here, we can see that we have applied the `pymethods` macro to our `impl` block. We also apply the `staticmethod` macro to our `process_numbers` static method. This function was used before, in the previous section, to process lists of lists. Now that our static method is defined, we can use this in our constructor method in the next step.

Defining a class constructor

Here are the steps we need to take:

1. We can define our constructor method in our `impl` block by running the following code:

```
#[new]
fn new(number: Vec<i32>, numbers: Vec<Vec<i32>>) \
    -> Self {
    let input_numbers: Vec<Vec<i32>> = \
        numbers.clone();
    let input_number: Vec<i32> = number.clone();

    let number_results: Vec<u64> =
        input_number.iter(
                            ).map(
            |x| fibonacci_number(*x)
        ).collect();

    let numbers_results: Vec<Vec<u64>> = Self::
                process_numbers(input_numbers);
        return FibProcessor {number, numbers,
                number_results, numbers_results}
}
```

Here, we accept inputs for the calculations of the Fibonacci numbers. We then clone them because we are going to pass them through the Fibonacci number functions. Once this is done, we apply the `fibonacci_number` function by mapping the input and collecting the results. We also collect the results from our static method. Once all the data is calculated, we construct the class and return it. Once this is done, all we must do is connect our class to our module.

2. This can be done by publicly declaring our class file in the `src/class_module/mod.rs` file, as follows:

```
pub mod fib_processor;
```

3. Now that this is done, we import it into our `src/lib.rs` file by running the following code:

```
mod class_module;
```

```
use class_module::fib_processor::FibProcessor;
```

4. Once this is done, we can add our class to our module in the same file, as follows:

```
m.add_class::<FibProcessor>()?;
```

We have now fully integrated our class into the `pip` module.

Wrapping up and testing our module

As always, when we get to the end of a section, we must remember to do the following:

*   Update the GitHub repository.
*   Uninstall the current `pip` module.
*   Reinstall it in our Python environment.

Now that we have finished building our module and updated the installed version, we can manually test our module in the Python system by following these next steps:

1. We can open our Python shell and test our class by running the following code:

```
>>> from flitton_fib_rs.flitton_fib_rs import
FibProcessor
>>> test = FibProcessor([11, 12, 13, 14], [[11, 12],
                               [13, 14], [15, 16]])
>>> test.numbers_results
[[89, 144], [233, 377], [610, 987]]
```

2. We can see that our Rust object works seamlessly in our Python system with calculated results. We must remember that we have set rules around our attributes. To check this, we can try to assign our `results` attribute, which will give us the following output:

```
>>> test.numbers_results = "test"
Traceback (most recent call last):
  File "<stdin>", line 1, in <module>
```

```
AttributeError: attribute 'numbers_results' of
'builtins.FibProcessor' objects is not writable
```

3. Here, we can see that our `results` attribute is not writable. We can also test typing. Although our `number` attribute is writable, it is supposed to be a vector of integers. If we try to assign a string to this attribute, we get the following printout:

```
>>> test.number = "test"
Traceback (most recent call last):
  File "<stdin>", line 1, in <module>
TypeError: 'str' object cannot be interpreted as an
integer
```

4. Here, we can see that our typing is also enforced, even though it looks and acts like a native Python object. Finally, we can test to see if we can write a new value to the `number` attribute by running the following code:

```
>>> test.number = [1, 2, 3, 4, 5]
>>> test.number
[1, 2, 3, 4, 5]
```

It seems that we can write when the type and permissions are correct. Considering all of this, what is the point of creating these classes? They make the interface for our module smoother, but how much faster is this class?

To quantify this, we can create a simple testing Python script in our Python environment, as follows:

1. First, in our Python script, we import our Rust class and the `time` module by running the following code:

```
from flitton_fib_rs.flitton_fib_rs import FibProcessor
import time
```

2. We must now create a pure Python object with the same functionality in this script, as follows:

```
class PythonFibProcessor:

    def __init__(self, number, numbers):
        self.number = number
        self.numbers = numbers
        self.numbers_results = None
```

```python
            self.number_results = None
            self._process()

    def _process(self):
        self.numbers_results = \
            [self.calculate_numbers(i)\
                for i in self.numbers]
        self.number_results = \
            self.calculate_numbers(
                self.number)

    def fibonacci_number(self, number):
        if number < 0:
            return None
        elif number <= 2:
            return 1
        else:
            return self.fibonacci_number(number - 1) + \
                    self.fibonacci_number(number - 2)

    def calculate_numbers(self, numbers):
        return [self.fibonacci_number(i) for i in \
            numbers]
```

3. Now that our benchmark pure Python object is defined, we are now at the timing stage of the script, where we put the same inputs into both classes and test them with the following code:

```python
t_one = time.time()
test = FibProcessor([11, 12, 13, 14], [[11, 12], \
        [13, 14], [15, 16]])
t_two = time.time()
print(t_two - t_one)

t_one = time.time()
test = PythonFibProcessor([11, 12, 13, 14], \
        [[11, 12], [13, 14], [15, 16]])
```

```
t_two = time.time()
print(t_two - t_one)
```

4. Running this gives us the following output:

```
1.4781951904296875e-05
0.0007779598236083984
```

This translates to the following:

```
0.000017881393432617188
0.0007779598236083984
```

Remember, the Rust class is the top reading. This means that our Rust class is *43 times faster than our Python class!* To put this into perspective, we can see the difference in the following screenshot:

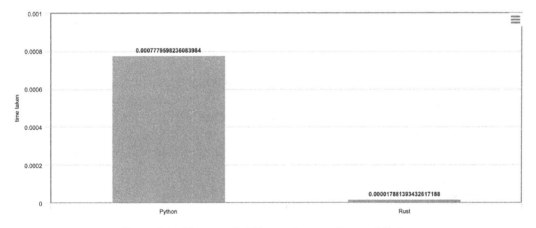

Figure 6.4 – Class speed difference between Rust and Python

Here, we can see that our class interfaces built in Rust are faster than our Python classes. `pyo3` supports class inheritance and other features. More resources on this are supplied in the *Further reading* section. We now have a strong base when it comes to working with Python objects in Rust. There are always more features to read up on, and these can be built on top of the structures that we have built.

Summary

In this chapter, we added a third-party `pip` module into our `setup.py` file so that we could add another entry point that could read .yml files. We read the .yml file and passed the data from that file in the form of a dictionary into our Rust functions, handling the complex data structure under the `PyDict` struct. We then downcasted data from our complex data structure into other Python objects and Rust data types. This gave us the power to handle a range of Python data types passed into our Rust code, giving us extra flexibility in how our Python code interacts with our Rust code.

We went one step further than complex Python data structures by accepting custom Python objects under the `PyAny` struct. Once we accepted custom Python objects, we could inspect attributes and set them as and when we wanted to. We even acquired the Python GIL to create our own Python data structures to help us work with the custom Python objects passed into our Rust code. To polish off our Python object skills, we built Python classes within our Rust code that not only can be imported into the Python system, acting just like a pure Python class, but are also 44 times faster. We now have a powerful tool that will not only speed up our Python code but will also enable us to interact with Python systems seamlessly.

In the next chapter, we tackle the final hurdle that is stopping us from infusing Rust into every Python project we have. People reach for Python due to the extensive third-party modules that are built for it, such as statistical and **machine learning** (**ML**) packages. We will work with the third-party `numpy` module and use it in our Rust code. This will enable us to utilize third-party Python modules in our Rust extension.

Questions

1. How do you extract a vector of `i32` integers from a `PyDict` struct?

2. If we have a vector of strings but we apply a `.extract::<Vec<i32>>()` function on it and we directly unwrap it, what will happen?

3. How would you be able to loop through a `Vec<i32>` vector, doubling each item and packaging the results in another vector in one line of Rust code?

4. If we acquire the Python GIL to create a `PyDict` struct, will this affect the Python system in any way?

5. Although our Python classes built in our Rust code essentially run the same way as our pure Python classes, there are some core differences. What are they?

Answers

1. First, we must get a list from the `PyDict` struct by applying the `get_item` function to `PyDict`. If there is data under the key that we use, we then perform `.downcast::<PyList>()` to convert our data into a `PyList` struct. If we achieve this, we then perform `.extract::<Vec<i32>>()` on the `PyList` struct, giving us a `Vec<i32>`.

2. Our `extract` function will automatically throw a `PyTypeError` Python-friendly error.

3. With this, we use the `iter`, `map`, and `collect` functions, as follows:

```
let results: Vec<i32> = some_vector.iter().map(
        |x| 2*x
    ).collect();
```

4. No—the Python system that is running the code has already acquired the GIL. If it does not have the GIL, it would just wait for another thread to finish before acquiring the GIL.

5. The typing system is still enforced. If we try to set an attribute that is a list of integers to a string, an error will be thrown. Another difference is that `set` and `get` macros for each attribute must be defined. If they are not, then the attribute cannot be accessed or set.

Further reading

- *PyO3 (2021). PyO3 user guide—Python Classes* https://pyo3.rs/v0.13.2/class.html

7

Using Python Modules with Rust

We have now become comfortable with writing Python packages in Rust that can be installed using `pip`. However, a large advantage of Python is that it has a lot of mature Python libraries that help us write productive code with minimal errors. This seems a legitimate observation that could halt us from adopting Rust in our Python system. However, in this chapter, we counter this observation by importing Python modules into our Rust code and running Python code in our Rust code. To achieve an understanding of this, we are going to use the **NumPy** Python package to implement a basic mathematical model. Once this is done, we are going to use the NumPy package in our Rust code to simplify the implementation of our mathematical model. Finally, we will evaluate the speed of both implementations.

In this chapter, we will cover the following topics:

- Exploring NumPy

- Building a model in NumPy

- Using NumPy and other Python modules in Rust

- Recreating our NumPy model in Rust

After completing this chapter, we will be able to import Python packages into our Rust code and use it. This is powerful, as relying on a certain Python package would not hold us back from implementing Rust in our Python systems for a certain task. The solutions that we implement in this chapter using pure Python, Rust, and NumPy will also give us an understanding of the trade-offs of each implementation when it comes to code complexity and speed so that we do not try to implement a *one-size-fits-all* solution for every problem, avoiding sub-optimal solutions.

Technical requirements

The code for this chapter can be found via the following GitHub link:

```
https://github.com/PacktPublishing/Speed-up-your-Python-with-Rust/tree/main/chapter_seven
```

Exploring NumPy

Before we start using NumPy in our own modules, we must explore what NumPy is and how to use it. NumPy is a third-party computational Python package that enables us to perform calculations on lists. NumPy is mainly written in the C language, meaning that it will be faster than pure Python. In this section, we will have to assess whether our NumPy implementation beats a Rust implementation that is imported into Python.

Adding vectors in NumPy

NumPy enables us to build vectors that we can loop through and apply functions to. We can also perform operations between vectors. We can demonstrate the power of NumPy by adding items of each vector together, as seen here:

```
[0, 1, 2, 3, 4]
[0, 1, 2, 3, 4]
---------------
[0, 2, 4, 6, 8]
```

To achieve this, we initially need to import modules by running the following code:

```
import time
import numpy as np
import matplotlib.pyplot as plt
```

With this, we can build a `numpy_function` NumPy function that creates two NumPy vectors of a certain size and adds them together by running the code presented here:

```
def numpy_function(total_vector_size: int) -> float:
    t1 = time.time()
    first_vector = np.arange(total_vector_size)
    second_vector = np.arange(total_vector_size)
    sum_vector = first_vector + second_vector
    return time.time() - t1
```

Here, we can see that we can add the vectors by merely using the addition operator. Now that we have our function defined, we can plot how this scales by looping through a list of integers and applying `numpy_function` to the items, collecting the results in a list, by running the code shown here:

```
numpy_results = [numpy_function(i) for i in range(0 \
    10000)]
plt.plot(numpy_results, linestyle='dashdot')
plt.show()
```

This gives us a line plot, as shown here:

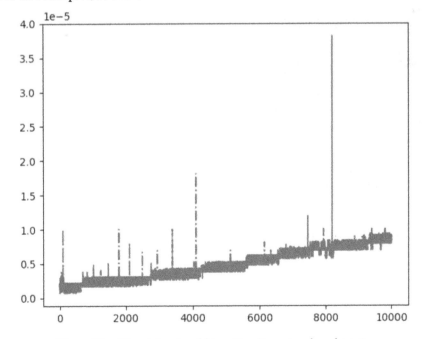

Figure 7.1 – Time taken to add two NumPy vectors based on size

We can see in the preceding screenshot that the increase is linear. This is expected because there is only one loop when adding each integer in the vector to the other vector. We can also see that there are points where the time taken shoots up—this is the garbage collection kicking in. To appreciate the effect NumPy has, we can redefineour example by adding both vectors with a list in pure Python in the next subsection.

Adding vectors in pure Python

We can add two vectors in pure Python and time this by running the following code:

```python
def python_function(total_vector_size: int) -> float:
    t1 = time.time()
    first_vector = range(total_vector_size)
    second_vector = range(total_vector_size)
    sum_vector = [first_vector[i] + second_vector[i] for \
      i in range(len(second_vector))]
    return time.time() - t1
```

With our new Python function, we can run both NumPy and Python functions and chart them by running the following code:

```python
print(python_function(1000))
print(numpy_function(1000))

python_results = [python_function(i) for i in range(0, \
    10000)]
numpy_results = [numpy_function(i) for i in range(0, \
    10000)]

plt.plot(python_results, linestyle='solid')
plt.plot(numpy_results, linestyle='dashdot')
plt.show()
```

This gives us the following results:

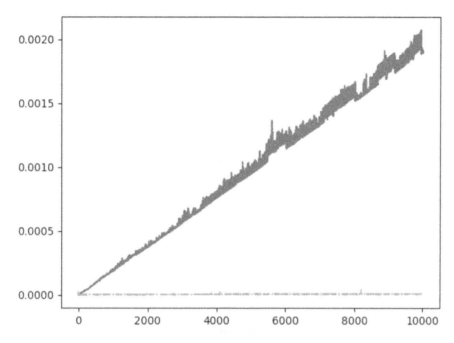

Figure 7.2 – Time taken to add two vectors based on size

As we can see in *Figure 7.2*, the NumPy vectors are represented in the bottom line and pure Python is represented in the increasing line, so we can conclude that Python does not scale that well when compared to our NumPy implementation. The output makes it clear that NumPy is a good choice when performing calculations on big vectors. However, how does this compare to our Rust implementation? We explore this in the next subsection.

Adding vectors using NumPy in Rust

To compare how NumPy compares to our Rust implementation, we must incorporate an adding vector function to our Rust package that we have been building throughout this book so far. Here are the steps we need to take:

1. Considering that this is a test function that we are using for demonstrative purposes, we can merely insert it into our `lib.rs` file. All we do is build a `time_add_vectors` function that accepts a number, create two vectors of a size equal to the number passed as input, loop through them at the same time, and add the items together, as shown here:

```
#[pyfunction]
fn time_add_vectors(total_vector_size: i32)
```

```
        -> Vec<i32> {
    let mut buffer: Vec<i32> = Vec::new();
    let first_vector: Vec<i32> =
      (0..total_vector_size.clone()
        ).map(|x| x).collect();
    let second_vector: Vec<i32> = \
      (0..total_vector_size
        ).map(|x| x).collect();

    for i in &first_vector {
        buffer.push(first_vector[**&i as usize] +
                    second_vector[*i as usize]);
    }
    return buffer
}
```

2. Once we have done this, we must remember to add this function to our module, as follows:

```
#[pymodule]
fn flitton_fib_rs(_py: Python, m: &PyModule) -> \
  PyResult<()> {
    . . .
    m.add_wrapped(wrap_pyfunction!(time_add_vectors));
    . . .
    Ok(())
}
```

We must remember to update our GitHub repository and reinstall our Rust package in our Python environment.

3. Once this is done, we must implement the function in our Python testing script and time it. First, we must import it with the code shown here:

```
import time
import matplotlib.pyplot as plt
import numpy as np
from flitton_fib_rs import time_add_vectors
```

4. Once this is done, we can define our `rust_function` Python function that calls the `time_add_vectors` function and times how long it takes to complete the addition by running the following code:

```
def rust_function(total_vector_size: int) -> float:
    t1 = time.time()
    sum_vector = time_add_vectors(total_vector_size)
    result = time.time() - t1
    if result > 0.00001:
        result = 0.00001
    return result
```

You may have noticed that we trim `result` returned by `rust_function`. This is not cheating—we do this because when the garbage collector kicks in, it can cause spikes and ruin the scaling of the graph. We will also do this with our NumPy function by running the following code:

```
def numpy_function(total_vector_size: int) -> float:
    t1 = time.time()
    first_vector = np.arange(total_vector_size)
    second_vector = np.arange(total_vector_size)
    sum_vector = first_vector + second_vector
    result = time.time() - t1
    if result > 0.00001:
        result = 0.00001
    return result
```

We can see that we are applying the same metrics to both the functions so that one will not be artificially lowered compared to the other.

5. Now that we have done this, we need to define the pure function by running the following code:

```
def python_function(total_vector_size: int) -> float:
    t1 = time.time()
    first_vector = range(total_vector_size)
    second_vector = range(total_vector_size)
    sum_vector = [first_vector[i] + second_vector[i] for
        i in range(len(second_vector))]
    result = time.time() - t1
    if result > 0.0001:
```

```
        result = 0.0001
    return result
```

It must be noted that we have a higher cut-off for the pure Python because we expect the base reading to be much higher.

6. Now that we have all of our metric functions for Rust, NumPy, and pure Python, we can create Python lists with the results of both functions and plot them by running the following code:

```
numpy_results = [numpy_function(i) for i in range(0, \
    300)]
rust_results = [rust_function(i) for i in range(0, \
    300)]
python_results = [python_function(i) for i in range \
    (0,300)]


plt.plot(rust_results, linestyle='solid', \
    color="green")
plt.plot(python_results, linestyle='solid', \
  color="red")
plt.plot(numpy_results, linestyle='solid', \
color="blue")
plt.show()
```

Running our code will give us the result shown here:

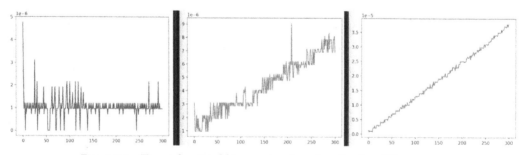

Figure 7.3 – Time taken to add two vectors based on the size of the vectors;
left: NumPy; middle: Rust; right: pure Python

In the preceding screenshot, we can see that NumPy is the fastest and is not scaling aggressively as the size of the vector increases. Our Rust implementation is a lot faster than our pure Python implementation by an order of magnitude, but it is not as efficient as NumPy.

We can see that Python optimizations such as NumPy do increase the speed compared to pure Python and Rust. However, NumPy's clean syntax of simply adding vectors is not the only functional advantage of this module. In the next section, we will explore another functionality that NumPy has that would require a lot of extra code if we were going to try to code it from scratch in Python or Rust.

Building a model in NumPy

In this section, we are going to build a basic mathematical model to demonstrate the power that NumPy has apart from speed. We are going to use matrices to make a simple model. To achieve this, we will have to carry out the following steps:

1. Define our model.
2. Build a Python object that executes our model.

Let's look at these steps in detail in the following subsections.

Defining our model

A mathematical model is essentially a set of weights that calculate an outcome based on inputs. Before we go any further, we must remember the scope of this book. We are building a model to demonstrate how to utilize NumPy. If we covered the nuances of mathematical modeling, that would take up the whole book. We will be building a model based on the example discussed in the previous section, but this does not mean that the model defined is an accurate description of the complexity of mathematical modeling. Here are the steps we need to take:

1. We start by looking at a very simple mathematical model that would be a simple speed equation, as shown here:

$$speed = \frac{distance}{time}$$

2. With our model, we can calculate the time taken to complete a journey with the rearrangement shown here:

$$time = \frac{distance}{speed} \rightarrow time = \frac{1}{speed} distance \rightarrow t = ax$$

The final equation on the right is merely substituting the values for letters so that they can be plugged into a bigger model without it taking up an entire page.

3. Now, let's take our model a little further. We collect some data from a trucking company, and we manage to quantify different grades of traffic into numbers and fit our data so that we can produce a weight to describe the effect traffic has on time. With this, our model has evolved to the one defined here:

$$t = \alpha x + \beta y$$

Here, β is the weight of traffic, and y is the grade of traffic. As we can see, if the grade of traffic increases, so does the time. Now, let's say that the model is different for cars and trucks. This gives us the following set of equations:

$$\alpha_c x + \beta_c y = t_c$$

$$\alpha_t x + \beta_t y = t_t$$

We can deduce from the equations that the distance (x) and traffic grade (y) are the same for both cars and trucks. This makes sense. While the weights could be different as cars could be affected by distance and traffic differently, as denoted in their weights, the input parameters are the same.

4. Considering this, the equation could be defined as the following matrix equation:

$$\begin{bmatrix} \alpha_c & \beta_c \\ \alpha_t & \beta_t \end{bmatrix} \begin{bmatrix} x \\ y \end{bmatrix} = \begin{bmatrix} t_c \\ t_t \end{bmatrix}$$

This might seem excessive right now, but there are advantages to this. Matrices have a range of functions that enable us to perform algebra on them. We will cover a few here so that we can understand how NumPy becomes invaluable to us when calculating this model.

5. To do this, we must acknowledge that matrix multiplication must occur in a certain order for it to work. Our model is essentially calculated by the notation shown here:

$$\begin{bmatrix} \alpha_c & \beta_c \\ \alpha_t & \beta_t \end{bmatrix} \begin{bmatrix} x \\ y \end{bmatrix} = \begin{bmatrix} \alpha_c x & \beta_c y \\ \alpha_t x & \beta_t y \end{bmatrix} = \begin{bmatrix} \alpha_c x + \beta_c y \\ \alpha_t x + \beta_t y \end{bmatrix}$$

6. Our $x\,y$ matrix must be on the right of our weights matrix. We can add more inputs to our $x\,y$ matrix with the following notation:

$$\begin{bmatrix} \alpha_c & \beta_c \\ \alpha_t & \beta_t \end{bmatrix} \begin{bmatrix} x_1 & x_2 & x_3 \\ y_1 & y_2 & y_3 \end{bmatrix} = \begin{bmatrix} t_{c1} & t_{c2} & t_{c3} \\ t_{t1} & t_{t2} & t_{t3} \end{bmatrix}$$

7. We can, in fact, keep stacking our inputs, and we will get proportional outputs. This is powerful; we can put in an input matrix of any size if we keep the dimensions of the matrix consistent. We can also invert our matrix. If we invert our matrix, we can then input times to work out the distance and grade of the traffic. Inverting a matrix takes the following form:

$$\begin{bmatrix} a & b \\ c & d \end{bmatrix}^{-1} = \frac{1}{ad - bc} \begin{bmatrix} d & -b \\ -c & a \end{bmatrix} = \begin{bmatrix} \dfrac{d}{ad - bc} & \dfrac{-b}{ad - bc} \\ \dfrac{-c}{ad - bc} & \dfrac{a}{ad - bc} \end{bmatrix}$$

8. Here, we can see that if we multiply a scalar by the matrix, it just gets applied to all elements of the matrix. Considering this, our model can calculate the traffic grade and distance using the inverse matrix with the following notation:

$$\begin{bmatrix} \alpha_c & \beta_c \\ \alpha_t & \beta_t \end{bmatrix}^{-1} \begin{bmatrix} t_c \\ t_t \end{bmatrix} = \begin{bmatrix} x \\ y \end{bmatrix}$$

We have only covered enough matrix mathematics to code our model, but even with just this, we can see that matrices enable us to manipulate multiple equations and shuffle them around to calculate different things quickly. However, if we were to code the matrix multiplications from scratch, it would take a lot of time and we would run the risk of performing errors. To have fast, safe development of our model, we will need to use NumPy module functions, which is what we will do in the next subsection.

Building a Python object that executes our model

We saw in the previous section that there are two different paths that we can take. When we build our model, we are going to have two branches—one for calculating the time taken, and the other for calculating the traffic and distance from time. To build our model class, we must map out our dependencies, as shown here:

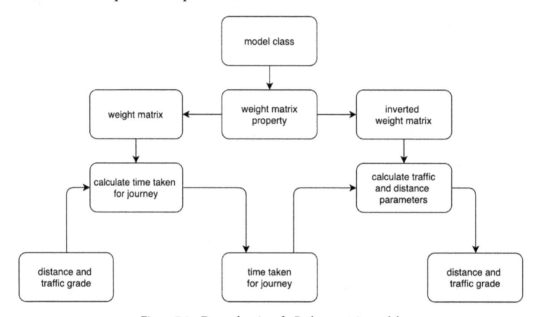

Figure 7.4 – Dependencies of a Python matrix model

The preceding diagram shows us that we must define the weight matrix property before anything else as this property is the main mechanism on which everything else is calculated. This was also evident in the matrix equation. We can build our class with the weight matrix property, as follows:

```python
import numpy as np


class MatrixModel:


    @property
    def weights_matrix(self) -> np.array:
        return np.array([
            [3, 2],
            [1, 4]
        ])
```

Here, we can see that we use NumPy for our matrix and that our matrix is a list of lists. We use NumPy arrays over normal arrays because NumPy arrays have matrix operations such as `transpose`. Remember that the positions of the matrices matter when they are being multiplied. For instance, we have a simple matrix equation as follows:

$$A \cdot B = \begin{bmatrix} 2 & 4 \end{bmatrix} \begin{bmatrix} 6 \\ 8 \end{bmatrix} = \begin{bmatrix} 2 \times 6 & 4 \times 8 \end{bmatrix} = 44$$

If we were to swap the matrix order, the matrices would not be able to multiply due to their shapes not being compatible; this is where the `transpose` operation comes in. A `transpose` function flips the matrix, enabling us to switch the order of the multiplication. We will not be using `transpose` in our model, but the Python commands in the terminal here show us how NumPy gives us this function out of the box:

```
>>> import numpy as np
>>> t = np.array([
                [3, 2],
                [1, 4]
            ])
>>> t.transpose()
array([[3, 1],
       [2, 4]])
>>> x = np.array([
                    [3],
                    [1]
                ])
>>> x.transpose()
array([[3, 1]])
```

Here, we can see that the matrices that we have built with NumPy arrays can change shape with ease. Now that we have established that we are building our matrices with NumPy arrays, we can build a function that will call the function that accepts the distance of the journey and the traffic grade for our `MatrixModel` class, as follows:

```
def calculate_times(self, distance: int, \
    traffic_grade: int) -> dict:
        inputs = np.array([
            [distance],
            [traffic_grade]
        ])
```

```
    result = np.dot(self.weights_matrix, inputs)
    return {
        "car time": result[0][0],
        "truck time": result[1][0]
    }
```

Here, we can see that once we have constructed our input matrix, we multiply this by our weights matrix with the np.dot function. result is a matrix, which—as we know—is a list of lists. We unpack this and then return it in the form of a dictionary.

We have nearly finished our model; all we must do now is now build our inverse model. This is where we pass in the times taken for the journey to calculate the distance and traffic grade for our MatrixModel class. This is done with the following code:

```
    def calculate_parameters(self, car_time: int,
                                   truck_time: int) -> dict:
        inputs = np.array([
            [car_time],
            [truck_time]
        ])
        result = np.dot(np.linalg.inv(self. \
          weights_matrix), inputs)
        return {
            "distance": result[0][0],
            "traffic grade": result[1][0]
        }
```

Here, we can see that we take the same approach; however, we use the np.linalg.inv function to get the inverse of the self.weights_matrix matrix. Now that this is done, we have a fully functioning model and we can test it, as follows:

```
test = MatrixModel()

times = test.calculate_times(distance=10, traffic_grade=3)
print(f"here are the times: {times}")

parameters = test.calculate_parameters(
    car_time=times["car time"], truck_time=times["truck \
      time"]
```

```
)
print(f"here are the parameters: {parameters}")
```

Running the preceding code will give us the following printout in the terminal:

```
{'car time': 36, 'truck time': 22}
{'distance': 10.0, 'traffic grade': 3.0}
```

With this terminal printout, we can see that our model works and that our inverse model returns the original inputs. With this, we can also conclude that NumPy is more than just speeding up our code; it gives us extra tools to solve problems such as modeling with matrices. This is the last hurdle that could stop us from reaching for Rust. In the next section, we will use the NumPy Python module in Rust by recreating our model in Rust.

Using NumPy and other Python modules in Rust

In this section, we are going to understand the basics of importing a Python module such as NumPy in our Rust program and return the result to our Python function. We will build our functionality in our Fibonacci number package that we have been coding so far in this book. We will also briefly explore importing a Python module in a generic sense so that you experience how to use a Python module that has the functionality you are relying on. We will build a more comprehensive approach to using Python modules in our Rust code in the next section. For this section, we will write all our code in the `src/lib.rs` file. Here are the steps we need to take:

1. First, we need to acknowledge that we pass in a dictionary and return the results in it. Because of this, we must import the `PyDict` struct by running the following code:

    ```
    use pyo3::types::PyDict;
    ```

2. Now that this is imported, we can define our function by running the following code:

    ```
    #[pyfunction]
    fn test_numpy<'a>(result_dict: &'a PyDict)
                -> PyResult<&'a PyDict> {
        let gil = Python::acquire_gil();
        let py = gil.python();
        let locals = PyDict::new(py);
        locals.set_item("np",
    ```

```
            py.import("numpy").unwrap());
    }
```

Because we are using a Python module, there is no surprise that we acquire the **global interpreter lock (GIL)** and get Python to interact with Python objects inside our Rust code. It must be noted that we also create a PyDict struct called locals.

We then import the NumPy module using the py.import function, inserting it into our localsstruct.

As demonstrated here, we will be using our locals struct as Python storage:

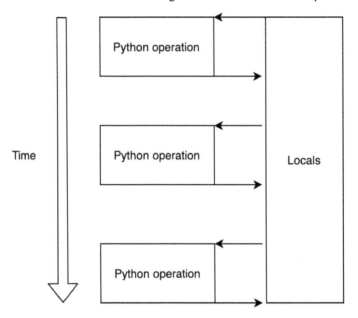

Figure 7.5 – Rust flow for computing Python processes within Rust

Here, every time we run a Python operation in our Rust code, we will pass Python objects from the locals into the Python computation. We then pass any new Python variables we need to add to our PyDict locals struct.

3. Now that we understand the flow, we can compute our first Python computation inside our test_numpy function by running the following code:

```
let code = "np.array([[3, 2], [1, 4]])";
let weights_matrix = py.eval(code,
                             None,
                             Some(&locals)).unwrap();
locals.set_item("weights_matrix", weights_matrix);
```

Here, we can see that we define the Python command as a string literal. We then pass this into our `py.eval` function. Our `None` parameter is for global variables. We are going to refrain from passing in global variables to keep this simple. We also pass in our `PyDict` `locals` struct to get the NumPy module we imported under the `np` namespace. We then unwrap the result and add this to our `localsstruct`.

4. We can now create an input NumPy vector and insert the outcome into our `localsstruct` by running the following code:

```
let new_code = "np.array([[10], [20]])";
let input_matrix = py.eval(new_code, None,
                        Some(&locals)).unwrap();
locals.set_item("input_matrix", input_matrix);
```

5. Now that we have both of our matricies in our `locals` storage, we can multiply them together, add them to our input dictionary, and return the result by running the following code:

```
let calc_code = "np.dot(weights_matrix, \
    input_matrix)";
let result_end = py.eval(calc_code, None,
                        Some(&locals)).unwrap();
result_dict.set_item("numpy result", result_end);
return Ok(result_dict)
```

With this, we can now use NumPy in our Rust code and get the results to pass them back into the Python system. We must remember to update our GitHub repository and reinstall our Rust package in our Python system. To test this, we can carry out the following console commands:

```
>>> from flitton_fib_rs import test_numpy
>>> outcome = test_numpy({})
>>> outcome["numpy result"].transpose()
array([[70, 90]])
```

Here, we can see that our NumPy process works inside Rust and returns Python objects that we can use just like all other Python objects. We could have done this using the Rust NumPy module, which gives us NumPy Rust structs. However, with the approach that we have covered, nothing is stopping us from using any Python module that we wish. We now have a full tool belt for fusing Python with Rust. In the next section, we will structure our NumPy model in Rust over a range of functions so that we can put in times for the inverse calculation and grade traffic with distance to calculate the times.

Recreating our NumPy model in Rust

Now that we can use our NumPy module in Rust, we need to explore how to structure it so that we can use Python modules to solve bigger problems. We will do this by building a NumPy model with a Python interface. To achieve this, we can break down the processes into functions that can be used as and when we need them. The structure of our NumPy model can be seen here:

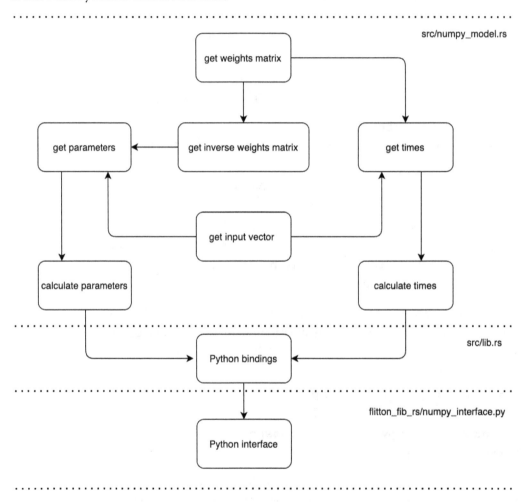

Figure 7.6 – Rust NumPy model structure

Considering the flow of our model structure in the preceding diagram, we can build our NumPy model in Rust with the following steps:

1. Build `get_weight_matrix` and `inverse_weight_matrix` functions.
2. Build `get_parameters`, `get_times`, and `get_input_vector` functions.
3. Build `calculate_parameters` and `calculate_times` functions.
4. Add calculate functions to the Python bindings and add a NumPy dependency to our `setup.py` file.
5. Build our Python interface.

We can see that each step has dependencies from the previous step. Let's have a detailed look at each of these steps in the following subsections.

Building get_weight_matrix and inverse_weight_matrix functions

Our weight and inverse weight matrices enable us to calculate the times and then recalculate the parameters inputted based on those times. We can start building our weight matrix function in the `src/numpy_model.rs` file with the following steps:

1. Before we write any code, we can import what we need by running the following code:

    ```
    use pyo3::prelude::*;
    use pyo3::types::PyDict;
    ```

    We will be using the `PyDict` struct to pass data between our functions and `pyo3` macros to wrap the functions and get the Python GIL.

2. Now that we have all of our imports, we can build our weight matrix function by running the following code:

    ```
    fn get_weight_matrix(py: &Python, locals: &PyDict) \
        -> () {
        let code: &str = "np.array([[3, 2], [1, 4]])";
        let weights_matrix = py.eval(code, None,
                              Some(&locals)).unwrap();
        locals.set_item("weights_matrix", weights_matrix);
    }
    ```

Here, we can see that we accept a reference to Python and `locals` storage. With this, we run our code and add it to our `locals` storage. We do not have to return anything because these are just referencing via borrowing. This means that the `py` and `locals` variables are not deleted when the scope of the variable has run its course. It also means that the `locals` storage will be updated with our `weights_matrix` function even though nothing is returned. We will be using this approach in most of our functions shown in *Figure 7.6*.

3. Now that we have our approach defined, we can create our inverse matrix function by running the following code:

```
fn invert_get_weight_matrix(py: &Python,
                            locals: &PyDict) -> () {
    let code: &str = "np.linalg.inv(weights_matrix)";
    let inverted_weights_matrix = py.eval(code, None,
                               Some(&locals)).unwrap();
    locals.set_item("inverted_weights_matrix",
                    inverted_weights_matrix);
}
```

Clearly, the `invert_get_weight_matrix` function cannot be run unless we run our `get_weight_matrix` function beforehand. We could make this more robust with a `get_item` check for `weights_matrix` in our `locals` storage and run the `get_weight_matrix` function if the weights matrix is not there, but this is not essential. We now have our weights functions defined, so we can move on to our next step of building our input vectors and calculation functions.

Building get_parameters, get_times, and get_input_vector functions

Just as with the previous steps. we are going to get our parameters, times, and inputs by using three functions. We will also have to pass the Python struct and `locals` storage into these functions as they are also going to be using NumPy via Python. We define these three functions in the following steps:

1. Referring to *Figure 7.6*, we can see that our input vector function does not have any dependencies and the other two depend on the input vector. Considering this, we build our input vector function by running the following code:

```
fn get_input_vector(py: &Python, locals: &PyDict,
                    first: i32, second: i32) -> () {
```

```
        let code: String = format!("np.array([[{}], \
          [{}]])", first, second);
        let input_vector = py.eval(&code.as_str(), None,
                            Some(&locals)).unwrap();
        locals.set_item("input_vector", input_vector);
    }
```

Here, we can see that this vector is generic, so we can pass in the parameters or the times depending on the calculation that we need. We can see that we use the format! macro to pass our parameters into our Python code.

2. Now that our input vector function is defined, we can build our calculations by running the following code:

```
    fn get_times<'a>(py: &'a Python,
                     locals: &PyDict) -> &'a PyAny {
        let code: &str = "np.dot(weights_matrix, \
          input_vector)";
        let times = py.eval(code, None,
          Some(&locals)).unwrap();
        return times
    }


    fn get_parameters<'a>(py: &'a Python,
                          locals: &PyDict) -> &'a PyAny {
        let code: &str = "
        np.dot(inverted_weights_matrix, input_vector)";
        let parameters = py.eval(code, None,
                          Some(&locals)).unwrap();
        return parameters
    }
```

With the aforementioned functions, we can get the variables that we need and put them into our Python code that uses the NumPy np.dot function. We then return the result as opposed to adding it to locals. We do not need to add it to locals because we are not going to use the results in any other computations in Rust. Now that all the computation steps have been done, we can move on to our next step—building the calculation functions that run and organize the whole process.

Building calculate_parameters and calculate_times functions

With these calculation functions, we need to take in some parameters, get the Python GIL, define our `locals` storage, and then run a series of computation processes to get what we need. We can define a `calculate_times` function by running the following code:

```
#[pyfunction]
pub fn calculate_times<'a>(result_dict: &'a PyDict,
    distance: i32, traffic_grade: i32) -> PyResult<&'a \
      PyDict> {
    let gil = Python::acquire_gil();
    let py = gil.python();
    let locals = PyDict::new(py);
    locals.set_item("np", py.import("numpy").unwrap());

    get_weight_matrix(&py, locals);
    get_input_vector(&py, locals, distance, traffic_grade);
    result_dict.set_item("times", get_times(&py, locals));
    return Ok(result_dict)
}
```

Here, we can see that we get the weight matrix, then the input vector, and then insert the results into a blank `PyDict` struct and return it. We can see the flexibility in this approach. We can slot functions in and out whenever we want, and rearranging the order is not a struggle. Now that we have built our `calculate_times` function, we can build our `calculate_parameters` function by running the following code:

```
#[pyfunction]
pub fn calculate_parameters<'a>(result_dict: &'a PyDict,
    car_time: i32, truck_time: i32) -> PyResult<&'a PyDict> {
    let gil = Python::acquire_gil();
    let py = gil.python();
    let locals = PyDict::new(py);
    locals.set_item("np", py.import("numpy").unwrap());

    get_weight_matrix(&py, locals);
    invert_get_weight_matrix(&py, locals);
    get_input_vector(&py, locals, car_time, truck_time);
```

```
    result_dict.set_item("parameters",
        get_parameters(&py, locals));
    return Ok(result_dict)
}
```

We can see that we use the same approach as our `calculate_times` function, using the invert weights instead. We could refactor this to reduce the repeated code, or we could enjoy the maximum flexibility of having the two functions isolated against each other. Our model is built now, so we can move to our next step where we add our calculation functions to our Python bindings.

Adding calculate functions to the Python bindings and adding a NumPy dependency to our setup.py file

Now that we have all the model code needed to calculate parameters through two functions, we are going to have to enable our outside user to utilize these functions with the following steps:

1. In our `src/lib.rs` file, we must define our module by running the following code:

```
mod numpy_model;
```

2. Now that this module has been declared, we can import the functions by running the following code:

```
use numpy_model::__pyo3_get_function_calculate_times;
use numpy_model::__pyo3_get_function_calculate_ \
    parameters;
```

3. We then wrap our functions in our module by running the following code:

```
#[pymodule]
fn flitton_fib_rs(_py: Python, m: &PyModule) -> \
    PyResult<()> {

    . . .

    m.add_wrapped(wrap_pyfunction!(calculate_times));
        m.add_wrapped(wrap_pyfunction!(calculate_
parameters));

    . . .

}
```

Remember—. . . . denotes existing code. We now must accept that our Rust code has a dependency on NumPy, so in our `setup.py` file, our dependencies will look like this:

```
requirements=[
    "pyyaml>=3.13",
    "numpy"
]
```

At this point, there is nothing stopping us from using our NumPy model; however, it will be better with a simple Python interface, which we will define in the next step.

Building our Python interface

In the `src/numpy_model.rs` file, we import what we need and define a basic class by running the following code:

```python
from .flitton_fib_rs import calculate_times, \
calculate_parameters


class NumpyInterface:


    def __init__(self):
        self.inventory = {}
```

The `self.inventory` variable will be where we store the results. Our functions for our class should calculate the times and parameters by calling our Rust functions, as follows:

```python
    def calc_times(self, distance, traffic_grade):
        result = calculate_times({}, distance,
                                 traffic_grade)
        self.inventory["car time"] = result["times"][0][0]
        self.inventory["truck time"] = \
          result["times"][1][0]


    def calc_parameters(self, car_time, truck_time):
        result = calculate_parameters({}, car_time,
            truck_time)
        self.inventory["distance"] =
                        result["parameters"][0][0]
```

```
        self.inventory["traffic grade"] =
                        result["parameters"][1][0]
```

Now that our Python interface is built, we have finished our NumPy model.

We must remember to update our GitHub repository and reinstall our module. Once this is done, we can run the following Python console commands:

```
>>> from flitton_fib_rs.numpy_interface import
NumpyInterface
>>> test = NumpyInterface()
>>> test.calc_times(10, 20)
>>> test.calc_parameters(70, 90)
>>> test.inventory
{'car time': 70, 'truck time': 90,
 'distance': 9.999999999999998,
 'traffic grade': 20.0}
```

While this demonstrates how we can use Python modules within Rust, we have to be careful when to use them. For our NumPy model example, it would have just been better to use NumPy within our Python code. To be honest, there is not that much that you can do with Python modules that you cannot do in Rust. Rust already has a NumPy crate that we can use. We should be using the Python modules in the initial stage if we cannot find—or do not have time to find and learn—a Rust alternative module; however, over time, these should be phased out of your Rust code.

Summary

In this chapter, we completed our tool belt when it comes to building Python extensions in Rust by using Python modules in our Rust code. We got a deeper appreciation for modules such as NumPy by exploring matrix mathematics to create a simple mathematical model. This showed us that we use modules such as NumPy for other functionality such as matrix multiplication, as opposed to just using NumPy for speed. This was demonstrated when we manipulated multiple mathematical equations with a few lines of NumPy code and matrix logic.

We then used matrix NumPy multiplication functions in our Rust code to recreate our mathematical model using a flexible functional programming approach. We finished this off by making our interface in a Python class. We also must remember that the NumPy implementation was faster than our Rust code. This is partly down to poor implementation on our part and the C optimization in NumPy. This has shown us that while Rust is a lot faster than Python, solving problems with Python packages such as NumPy might still be faster until equivalent crates are coded in Rust.

We used a generic approach to using Python modules in Rust. Because of this, we can theoretically use any Python module that we want. This means that if the Python module that you are rewriting relies on the functionality of third-party Python modules such as NumPy, we are now able to create Rust functions that use them. Considering this, there is no generic technical hurdle stopping you from rewriting Python code in Rust and slotting it into your Python system.

In the next chapter, we will put everything that we have learned so far together to build a new Python package written in Rust from start to end.

Questions

1. What are the steps we must follow to run a Python module in Rust?

2. How do you import a Python module into your Rust code?

3. If we wanted to use our Python code result inside Rust, how would we do this?

4. When you compare speed graphs of Python/NumPy with Rust, the Python/NumPy code has a lot of spikes. What could be causing this?

5. Do you think our NumPy implementation in Rust will be slower or faster than calling NumPy from Python, and why?

Answers

1. We initially must get Python from the GIL. We then must build a `PyDict` struct
    in order to store and pass Python variables between Python executions. We then
    define the Python code as a string literal and pass this into our `py.eval` function
    with our `PyDict` storage.

2. We must make sure that we get Python from the GIL. We then use this to run the
    `py.eval` function with the import line of code passed in as a string literal.
    We must remember to pass in our `PyDict` storage to ensure that we can reference
    the module in the future.

3. We must remember that Python code returns a `PyAny` struct, which we can extract
    using the following code:

    ```
    let code = "5 + 6";
    let result = py.eval(code, None, Some(&locals)).unwrap();
    let number = result.extract::<i32>().unwrap();
    ```

    We can see that `number` should be `11`.

4. This is because the Python versions must keep stopping to clean up variables
    with the garbage collection mechanism.

5. It would be slightly slower. This is because we are essentially still running Python
    code but through an extra layer which is Rust.Considering this, we should be using
    Python code out of convenience as opposed
    to optimization.

Further reading

- NumPy documentation for Rust (2021): *Crate numpy:* `https://docs.rs/numpy/0.14.1/numpy/`

- Giuseppe Ciaburro (2020): *Hands-on Simulation Modeling with Python: Develop
  simulation models to get accurate results and enhance decision-making processes.*
  Packt Publishing.

8

Structuring an End-to-End Python Package in Rust

Now that we have covered enough Rust and pyo3 to theoretically build a range of real-world solutions, we must be careful. It would not be good if we decided to reinvent the wheel in Rust and ended up with a slower outcome after coding the solution. Hence, understanding how to solve a problem and testing our implementation is important. In this chapter, we will be building a Python package using Rust that solves a simplified real-world problem and loads data from files to build a catastrophe model. We will structure the package in a manner where we can slot in extra functionality if our model gets more complex. Once we build our model, we will test it to see whether our implementation is worth it in terms of scaling and speed.

In this chapter, we will cover the following topics:

- Breaking down a catastrophe modeling problem for our package
- Building an end-to-end solution as a package
- Utilizing and testing our package

This chapter enables us to take what we have learned throughout the book and solve a real-world problem and handle data files. Testing our solution will also enable us to avoid spending too much time on a solution that will have a slower result, preventing us from potentially missing our shot at implementing Rust in Python systems at our place of work.

Technical requirements

The code and data for this chapter can be found at `https://github.com/PacktPublishing/Speed-up-your-Python-with-Rust/tree/main/chapter_eight`.

Breaking down a catastrophe modeling problem for our package

The project that we are going to build is a catastrophe model. This is where we calculate the probability of a catastrophe such as a hurricane, flood, or terror attack happening in a particular geographical location. We could do this using longitude and latitude coordinates. However, if we are going to do this, it is going to take a lot of computational power and time with little benefit. For instance, if we were going to calculate the probability of the flooding at Charing Cross Hospital in London, we could use the coordinates *51.4869° N, 0.2195° W*.

However, if we use the coordinates *51.4865° N, 0.2190° W*, we would still be hitting Charing Cross Hospital, despite us changing the coordinates by *0.0004° N, 0.0005° W*. We could change the coordinates even more and we would still be hitting Charing Cross Hospital. Therefore, we would be doing loads of computations to calculate repeatedly the probability of flooding of the same building, which is not efficient. To combat this, we can break down the locations into bins and give them a numerical value, as shown here:

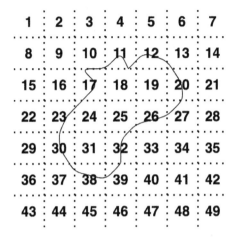

Figure 8.1 – Geographical bins for a catastrophe model of an island

Here, we can see that if a line of data in our model referred to bin 25, this means that the line of data is referring to land in the middle of our island that we are concerned with. We can make our calculations even more efficient. For instance, we can see that the squares in *Figure 8.1* with the coordinates of 33, 35, 47, and 49 and 1, 2, 8, and 9 are in the sea. Therefore, the probability of flooding in these squares is zero because it is already water, and there is nothing that we care about in terms of flooding in the sea. Because we are merely mapping our calculations onto these bins, nothing is stopping us from redefining all of the bins inside these squares as one bin.

Therefore, we must perform only one operation to calculate the risk of flooding in all our sea bins and that would be zero because the sea is already flooded. In fact, nothing is stopping us from sticking to square classifications for one bin. Bin number 1 could be all the squares that are 100% inside the sea, saving us a lot of time. We can also go the other way. We can make some of our bins more refined. For instance, areas near the coast might have more nuanced gradients of flooding, as a small distance closer to the sea could greatly increase the risk of flooding; therefore, we could break bin number 26 down into smaller bins. To avoid being dragged into the weeds, we will just refer to arbitrary bin numbers in our model data. Catastrophe modeling is its own subject, and we are merely using it to show how to build Rust Python packages that can solve real problems as opposed to trying to build the most accurate catastrophe model. Now that we understand how we map geographical data with probabilities, we can move on to the calculation of those probabilities.

Like with the mapping of geographical data, probability calculations are more complex and nuanced than what we are going to cover in this book. Companies like OASISLMF work with academic departments at universities to model risks of catastrophes and the damage inflicted. However, there is an overarching theme that we must do when calculating these probabilities. We will have to calculate the total probability of damage using the probability of the event happening in the area, and the probability of the event causing damage. To do this, we must multiply these probabilities together. We also must break down the probability of the event happening at a certain intensity. For instance, a category one hurricane is less likely to cause damage to a building compared to a category five hurricane. Therefore, we are going to run these probability calculations for each intensity bin.

We cannot go any further in designing our process without looking at the data that we have available. The data is in the form of CSV files and is available in our GitHub repository stated in the *Technical requirements* section. The first data file that we can inspect is the footprint.csv file. This file presents the probability of a catastrophe with a certain intensity happening in an area:

event_id	areaperil_id	intensity_bin_id	Probability
1	10	1	0.47
1	10	2	0.53

Here, we can see that we have taken in a series of event IDs. We can merge the footprint.csv data with the event IDs we passed in. This enables us to map the event IDs that we passed in with an area, intensity, and probability of it happening.

Now that we have merged our geographical data, we can now look at our damage data in the vulnerability.csv file:

vulnerability_id	intensity_bin_id	damage_bin_id	probability
1	1	1	0.45
1	2	2	0.65

Looking at this, we can merge the damage data of the intensity bin ID, duplicating whatever we need. We then must multiply the probabilities to get the total probability. The flow can be summed up as follows:

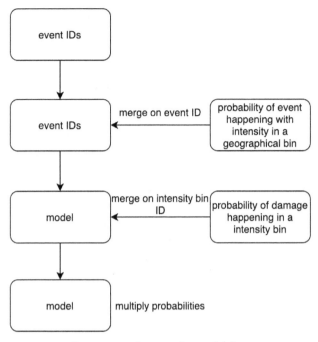

Figure 8.2 – Catastrophe model flow

Considering the data and flow, we can see that we now have events that have an intensity bin ID, damage bin ID, probability of the event happening in the area, and the probability of the event causing damage in a certain bin. These can then be passed on to another stage, which is the process of calculating financial losses. We will stop here, but we must remember that real-world applications need to adapt for expansion. For instance, there is interpolation. This is where we use a function to estimate the values across a bin, which is demonstrated here:

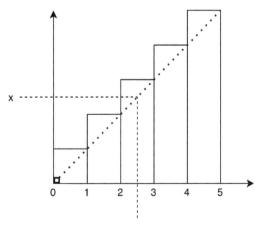

Figure 8.3 – Linear interpolation of a distribution

Here, we can see that if we just use bins, our reading between 2 and 2.9 would be the same. We know that the distribution is increasing, so we use a simple linear function, and the value of our reading increases as the reading increases. There are other more complex functions we can use, but this can increase the accuracy of readings if the bins are too wide. While we will not be using interpolation in our example, it is a legitimate step that we might want to slot in later. Considering this, our processes need to be isolated.

There is only one other thing that we must consider when designing our package, which is the storage of our model data. Our probabilities will be defined by an academic team that collected and analyzed a range of data sources and specific knowledge. For instance, damage to buildings requires structural engineering knowledge and knowledge of hurricanes. While we might expect our teams to update the models in later releases, we do not want the end user to easily manipulate data. We also do not want to hardcode the data into our Rust code; therefore, storing CSV files in our package would be useful for this demonstration. Considering this, our package should take the following structure:

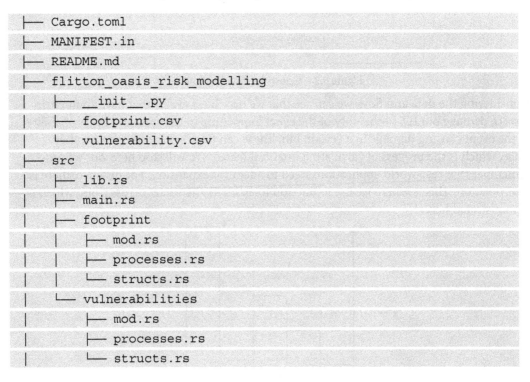

```
├── Cargo.toml
├── MANIFEST.in
├── README.md
├── flitton_oasis_risk_modelling
│   ├── __init__.py
│   ├── footprint.csv
│   └── vulnerability.csv
├── src
│   ├── lib.rs
│   ├── main.rs
│   ├── footprint
│   │   ├── mod.rs
│   │   ├── processes.rs
│   │   └── structs.rs
│   └── vulnerabilities
│       ├── mod.rs
│       ├── processes.rs
│       └── structs.rs
```

The structure should be familiar to you. In the preceding file structure, we can see that our merge processes for the probability of the event happening and the damage are in their own folders. Data structures for the process are housed in the `structs.rs` file and functions around the process are defined in the `processes.rs` file. The `flitton_oasis_risk_modelling` folder will house our compiled Rust code; therefore, our CSV files are also stored there.

We state that we are storing our CSV files in the `MANIFEST.in` file. Our `lib.rs` file is where our interface between our Rust and Python is defined. Now that we have defined the process for our catastrophe model, we can move on to the next section of building our end-to-end package.

Building an end-to-end solution as a package

In the previous section, we identified what we needed to do to build our catastrophe model package. We can achieve it with the following steps:

1. Build a footprint merging process.
2. Build a vulnerability and probability merging process.
3. Build a Python interface in Rust.
4. Build an interface in Python.
5. Build package installation instructions.

Before we build anything, we must define our dependencies in our `Cargo.toml` file with the following code:

```
[package]
name = "flitton_oasis_risk_modelling"
version = "0.1.0"
authors = ["Maxwell Flitton <maxwellflitton@gmail.com>"]
edition = "2018"

[dependencies]
csv = "1.1"
serde = { version = "1", features = ["derive"] }

[lib]
name = "flitton_oasis_risk_modelling"
crate-type=["rlib", "cdylib"]
```

```
[dependencies.pyo3]
version = "0.13.2"
features = ["extension-module"]
```

Here, we can see that we are using the `csv` crate to load our data and the `serde` crate to serialize the data that we had loaded from the CSV file. With this approach, it is important that we start by coding the processes first. This enables us to know what we need when we get to building our interfaces. Considering this, we can start building our footprint merging process.

Building a footprint merging process

Our footprint merging process is essentially loading our footprint data and merging it with our input IDs. Once this is done, we then return the data to be fed into another process. We initially need to build our data structs before we build our processes, as our processes will need them. We can build our footprint struct in the `src/footprint/structs.rs` file with the following code:

```rust
use serde::Deserialize;


#[derive(Debug, Deserialize, Clone)]
pub struct FootPrint {
    pub event_id: i32,
    pub areaperil_id: i32,
    pub intensity_bin_id: i32,
    pub probability: f32
}
```

Here, we can see that we apply the `Deserialize` macro to the struct so that when we load data from the file, it can be directly loaded into our `FootPrint` struct. We will also want to clone our struct if similar multiple event IDs are being passed into our package.

Now that we have our struct, we can build our merging process in our `src/`
`footprint/processes.rs` file:

1. First, we have to define the imports we need with the following code:

```
use std::error::Error;
use std::fs::File;
use csv;


use super::structs::FootPrint;
```

   We must remember that we did not define our struct in the `src/footprint/`
   `mod.rs` file, so this will not run yet, but we will define it in time before running
   our code.

2. We can now build a function that will read a footprint from the file with the
   following code:

```
pub fn read_footprint(mut base_path: String) -> \
  Result<Vec<FootPrint>, Box<dyn Error>> {
    base_path.push_str("/footprint.csv");
    let file = File::open(base_path.as_str())?;
    let mut rdr = csv::Reader::from_reader(file);


    let mut buffer = Vec::new();


    for result in rdr.deserialize() {
        let record: FootPrint = result?;
        buffer.push(record);
    }
    Ok(buffer)
}
```

Here, we can see our function requires the directory where our data file is housed.
We then add the filename to the path, open the file, and pass it through the
`from_reader` function. We then define an empty vector and add the data that we
deserialize. We now have a vector of `FootPrint` structs, which we return.

3. Now that we have our `load data` function, we can now build our `merge footprints` function in the same file with the following code:

```
pub fn merge_footprint_with_events(event_ids: \
    Vec<i32>,
        footprints: Vec<FootPrint>) -> Vec<FootPrint> {
    let mut buffer = Vec::new();

    for event_id in event_ids {
        for footprint in &footprints {
            if footprint.event_id == event_id {
                buffer.push(footprint.clone());
            }
        }
    }
    return buffer
}
```

Here, we can see that we take in a vector of event IDs and a vector of `FootPrint` structs. We then loop through our event IDs. For each event, we then loop through all the `FootPrint` structs, adding the struct to our buffer if it matches the event ID. We then return the buffer meaning that we have merged all that we need. We do not need to code any more processes. To make them useful, we can build an interface in the `src/footprint/mod.rs` file.

4. So, we must import what we need with the following code:

```
pub mod structs;
pub mod processes;

use structs::FootPrint;
use processes::{merge_footprint_with_events, \
    read_footprint};
```

5. Now that we have imported all that we need, we can build our interface in the same file with the following code:

```
pub fn merge_event_ids_with_footprint(event_ids: \
    Vec<i32>,
        base_path: String) -> Vec<FootPrint> {
```

```
    let foot_prints = \
        read_footprint(base_path).unwrap();
    return merge_footprint_with_events(event_ids, \
        foot_prints)
}
```

Here, we merely accept the file path and event IDs and pass them through our processes, returning the results.

With this, our footprint processes are built, meaning that we can move on to the next step of building the vulnerability merge processes.

Building the vulnerability merge process

Now that we have merged our event IDs with our footprint data, we have a working map of the probabilities of certain events happening at certain intensities within a range of geographical locations. We can merge this with the probabilities of damage occurring due to the catastrophe by following these steps:

1. In this process, we must load the vulnerabilities and then merge them with our existing data. To facilitate this, we will have to build two structs – one for the data that is loaded from the file and another for the result after the merge. Because we are loading the data, we will need to use the serde crate. In our src/vulnerabilities/structs.rs file, we import it with the following code:

   ```
   use serde::Deserialize;
   ```

2. We then build our struct to load the file with the following code:

   ```
   #[derive(Debug, Deserialize, Clone)]
   pub struct Vulnerability {
       pub vulnerability_id: i32,
       pub intensity_bin_id: i32,
       pub damage_bin_id: i32,
       pub probability: f32
   }
   ```

We must note here that the probability of the data we are loading is labeled under the probability field. This is the same with our FootPrint struct. Because of this, we must rename the probability field to avoid clashes during the merge. We also need to calculate the total probability.

3. Considering this, our result after the merge takes the form of the following code:

```
#[derive(Debug, Deserialize, Clone)]
pub struct VulnerabilityFootPrint {
    pub vulnerability_id: i32,
    pub intensity_bin_id: i32,
    pub damage_bin_id: i32,
    pub damage_probability: f32,
    pub event_id: i32,
    pub areaperil_id: i32,
    pub footprint_probability: f32,
    pub total_probability: f32
}
```

With this, our structs are complete and we can build our processes in our `src/vulnerabilities/processes.rs` file. Here, we are going to have two functions, reading the vulnerabilities, and then merging them with our model:

1. First, we must import everything that we need with the following code:

```
use std::error::Error;
use std::fs::File;
use csv;

use crate::footprint::structs::FootPrint;
use super::structs::{Vulnerability, \
    VulnerabilityFootPrint};
```

Here, we can see that we are relying on the `FootPrint` struct from our `footprint` module.

2. Now that we have everything, we can build our first process, which is loading the data with the following code:

```
pub fn read_vulnerabilities(mut base_path: String) \
    -> Result<Vec<Vulnerability>, Box<dyn Error>> {
        base_path.push_str("/vulnerability.csv");
        let file = File::open(base_path.as_str())?;
        let mut rdr = csv::Reader::from_reader(file);

        let mut buffer = Vec::new();
```

```
        for result in rdr.deserialize() {
            let record: Vulnerability = result?;
            buffer.push(record);
        }
        Ok(buffer)
    }
```

Here, we can see that this is similar code to our loading process in our footprint module. Refactoring this into a generalized function would be a good exercise.

3. Now that we have our loading function, we can merge Vec<Vulnerability> with Vec<FootPrint> to get Vec<VulnerabilityFootPrint>. We can define the function with the following code:

```
pub fn merge_footprint_with_vulnerabilities(
    vulnerabilities: Vec<Vulnerability>,
    footprints: Vec<FootPrint>) -> \
    Vec<VulnerabilityFootPrint> {
    let mut buffer = Vec::new();

    for vulnerability in &vulnerabilities {
        for footprint in &footprints {
            if footprint.intensity_bin_id == \
                vulnerability
                    .intensity_bin_id {
                    . . .
            }
        }
    }
    return buffer
}
```

Here, we can see that we have a new vector called buffer, which is where the merged data will be stored in the . . . placeholder. We can see that we loop through the footprints for each vulnerability. If intensity_bin_id matches, we execute the code in the . . . placeholder, which is the following code:

```
buffer.push(VulnerabilityFootPrint{
    vulnerability_id: vulnerability.vulnerability_id,
```

```
        intensity_bin_id: vulnerability.intensity_bin_id,
        damage_bin_id: vulnerability.damage_bin_id,
        damage_probability: vulnerability.probability,
        event_id: footprint.event_id,
        areaperil_id: footprint.areaperil_id,
        footprint_probability: footprint.probability,
        total_probability: footprint.probability * \
            vulnerability.probability
                });
```

Here, we are merely mapping the correct values to the correct fields of our
VulnerabilityFootPrint struct. In the last field, we calculate the total
probability by multiplying the other probabilities together.

Our processes are finally done, so we move on to building our interface for this process in
our src/vulnerabilities/mod.rs file:

1. We first import what we need with the following code:

```
pub mod structs;
pub mod processes;


use structs::VulnerabilityFootPrint;
use processes::{merge_footprint_with_vulnerabilities \
    ,read_vulnerabilities};
use crate::footprint::structs::FootPrint;
```

With this, we can create a function that takes in a base path for the directory of
where our data files are and our footprint data.

2. We then pass them through both of our processes, loading and merging, and then
return our merged data with the following code:

```
pub fn merge_vulnerabilities_with_footprint( \
    footprint: Vec<FootPrint>, mut base_path: String) \
        -> Vec<VulnerabilityFootPrint> {
    let vulnerabilities = read_vulnerabilities( \
        base_path).unwrap();
    return merge_footprint_with_vulnerabilities( \
        vulnerabilities, footprint)
}
```

We have now built our two processes for constructing our data model. We can move on to our next step, which is building our Python interface in Rust.

Building a Python interface in Rust

The Python interface is defined in the `src/lib.rs` file, where we use the `pyo3` crate to get our Rust code to communicate with the Python system. Here are the steps:

1. First, we must import what we need with the following code:

```
use pyo3::prelude::*;
use pyo3::wrap_pyfunction;
use pyo3::types::PyDict;

mod footprint;
mod vulnerabilities;

use footprint::merge_event_ids_with_footprint;
use vulnerabilities::merge_vulnerabilities_with_
footprint;
use vulnerabilities::structs::VulnerabilityFootPrint;
```

    Here, we can see that we import what we need from the `pyo3` crate. We will be wrapping a `get_model` function with `wrap_pyfunction` and returning a list of `PyDict` structs. We also define the process modules, structs, and functions that we need to build our model.

2. We can then define our function with the following code:

```
#[pyfunction]
fn get_model<'a>(event_ids: Vec<i32>, \
    mut base_path: String, py: Python) -> Vec<&PyDict> {
    let footprints = merge_event_ids_with_footprint( \
        event_ids, base_path.clone());
    let model = merge_vulnerabilities_with_footprint \
        (footprints, base_path);

    let mut buffer = Vec::new();

    for i in model {
```

```
        . . .
    }
    return buffer
}
```

It must be noted that we accept a `Python` struct into our function. This is automatically filled. If we get the `Python` struct via the **Global Interpreter Lock (GIL)** as done in previous chapters, we will not be able to return them because the lifetime will finish at the end of the function. Because we take in the `Python` struct, we can return the Python structures that we create in the function using the `Python` struct that we took in.

3. In the . . . placeholder, we create a `PyDict` struct with all the data for the model row and push it to our buffer with the following code:

```
let placeholder = PyDict::new(py);
placeholder.set_item("vulnerability_id", \
    i.vulnerability_id);
placeholder.set_item("intensity_bin_id", \
    i.intensity_bin_id);
placeholder.set_item("damage_bin_id", \
    i.damage_bin_id);
placeholder.set_item("damage_probability",\
    i.damage_probability);
placeholder.set_item("event_id", \
    i.event_id);
placeholder.set_item("areaperil_id",\
    i.areaperil_id);
placeholder.set_item("footprint_probability", \
    i.footprint_probability);
placeholder.set_item("total_probability", \
    i.total_probability);
    buffer.push(placeholder);
```

Here, we can see that we can push different types to our `PyDict` struct and Rust does not care.

4. We can then wrap our function and define our module with the following code:

```
#[pymodule]
fn flitton_oasis_risk_modelling(_py: Python, \
    m: &PyModule) -> PyResult<()> {
    m.add_wrapped(wrap_pyfunction!(get_model));
    Ok(())
}
```

Now that all our Rust programming is done, we can move on to building our Python interface in the next step.

Building our interface in Python

When it comes to our Python interface, we will have to build a function in a Python script in the `flitton_oasis_rist_modelling/__init__.py` file. We also store our data CSV files in the `flitton_oasis_rist_modelling` directory. Remember, we do not want our users interfering with the CSV files or having to know where they are. To do this, we will use the os Python module to find the directory of our module to load our CSV data.

To do this, we import what we need in the `flitton_oasis_rist_modelling/__init__.py` file with the following code:

```
import os
from .flitton_oasis_risk_modelling import *
```

Remember, our Rust code will compile into a binary and be stored in the `flitton_oasis_rist_modelling` directory, so we can do a relative import for all the wrapped functions in our Rust code. Now, we can code our `construct_model` model function with the following code:

```
def construct_model(event_ids):
    dir_path = os.path.dirname(os.path.realpath(__file__))
    return get_model(event_ids, str(dir_path))
```

Here, we can see that all the user needs to do is pass in the event IDs. However, if we tried to install this package using `pip`, we would get errors stating that the CSV files cannot be found; this is because our setup does not include the data files. We can solve this in our next step of building package installation instructions.

Building package installation instructions

To do this, we must state that we want to keep all CSV files in our MANIFEST.in file with the following code:

```
recursive-include flitton_oasis_risk_modelling/*.csv
```

Now that we have done this, we can move to our setup.py file to define our setup:

1. First, we must import what we need with the following code:

```python
#!/usr/bin/env python
from setuptools import dist
dist.Distribution().fetch_build_eggs([ \
    'setuptools_rust'])
from setuptools import setup
from setuptools_rust import Binding, RustExtension
```

Here, as we have done before, we fetch the setuptools_rust package; although it is not essential for the running of the package, it is needed for the installation.

2. We can now define our setup parameters with the following code:

```python
setup(
    name="flitton-oasis-risk-modelling",
    version="0.1",
    rust_extensions=[RustExtension(
    ".flitton_oasis_risk_modelling.flitton_oasis \
     _risk_modelling",
        path="Cargo.toml", binding=Binding.PyO3)],
    packages=["flitton_oasis_risk_modelling"],
    include_package_data=True,
    package_data={'': ['*.csv']},
    zip_safe=False,
)
```

Here, we can see that we do not need any Python third-party packages. We have also defined our Rust extension, set the include_package_data parameter to True, and defined our package data with package_data={'': ['*.csv']}. With this, all CSV files will be kept when installing our package.

3. We are nearly finished; all we have to do is define the `rustflags` environment variables in the `.cargo/config` file with the following code:

```
[target.x86_64-apple-darwin]
rustflags = [
    "-C", "link-arg=-undefined",
    "-C", "link-arg=dynamic_lookup",
]
[target.aarch64-apple-darwin]
rustflags = [
    "-C", "link-arg=-undefined",
    "-C", "link-arg=dynamic_lookup",
]
```

With this, we can upload our code and install it in our Python system.

We can now use our Python module. We can test this in our module with the terminal output, as follows:

```
>>> from flitton_oasis_risk_modelling import
  construct_model
>>> construct_model([1, 2])
[{'vulnerability_id': 1, 'intensity_bin_id': 1,
'damage_bin_id': 1, 'damage_probability': 0.44999998807907104,
'event_id': 1, 'areaperil_id': 10,
'footprint_probability': 0.4699999988079071,
'total_probability': 0.21149998903274536},
{'vulnerability_id': 1, 'intensity_bin_id': 1,
'damage_bin_id': 1, 'damage_probability':
0.44999998807907104,
'event_id': 2, 'areaperil_id': 20,
'footprint_probability': 0.30000001192092896,
'total_probability': 0.13500000536441803},
{'vulnerability_id': 1, 'intensity_bin_id': 2,
'damage_bin_id': 2, 'damage_probability':
0.6499999761581421,
'event_id': 1, 'areaperil_id': 10,
'footprint_probability': 0.5299999713897705,
'total_probability': 0.3444997544288635},
```

```
{'vulnerability_id': 1, 'intensity_bin_id': 2,
'damage_bin_id': 2,
'damage_probability': 0.6499999761581421, 'event_id': 2,
'areaperil_id': 20, 'footprint_probability':
0.699999988079071,
'total_probability': 0.45499998331069946},
. . .
```

There's more data that is printed out, but if your printout correlates with the preceding output, then there is a high chance that the rest of your data is accurate. Here, we have built a real-world solution that loads data and does a series of operations and processes to come up with a model. However, it is a basic model that would not be used in real-life catastrophe modeling; we have coded it in isolated modules so that we can slot in more processes when we need to.

However, we need to ensure that all our effort was not for nothing. We can do what we did in this package with a few lines of Python code using pandas, which is written in C, so it could be quicker or at the same speed. Considering this, we need to test to ensure that we are not wasting our time by testing our code in the next section.

Utilizing and testing our package

We have started building out our solution in a Python package coded in Rust. However, we need to justify to our team and ourselves that all this effort was worth it. We can test to see whether we should continue with our efforts in a single isolated Python script. In this Python script, we can test by following these steps:

1. Build a Python construct model using pandas.
2. Build random event ID generator functions.
3. Time our Python and Rust implementations with a series of different data sizes.

Once we have carried out all the aforementioned steps, we will know whether we should progress further with our module.

In our testing script, before we start coding anything, we must import all of what we need with the following code:

```
import random
import time

import matplotlib.pyplot as plt
```

```
import pandas as pd
from flitton_oasis_risk_modelling import construct_model
```

Here, we are using the `random` module to generate random event IDs and the `time` module to time our implementations. We are using `pandas` to build our model, `matplotlib` to plot the outcomes, and our Rust implementation. We can now build our model.

Building a Python construct model using pandas

Now that we have imported everything that we need, we can move on to loading data from the CSV files in Python and use it to construct a model in Python using pandas with the following steps:

1. First, our function must take in event IDs. We also must load our data from our CSV files with the following code:

```
def python_construct_model(event_ids):
    vulnerabilities = \
        pd.read_csv("./vulnerability.csv")
    foot_print = pd.read_csv("./footprint.csv")
    event_ids = pd.DataFrame(event_ids)
```

2. Now that we have all our data, we can merge our data and rename the `probability` column to avoid clashing with the following code:

```
    model = pd.merge(
        event_ids, foot_print, how="inner", \
            on="event_id"
    )
    model.rename(
        columns={"probability": \
            "footprint_probability"},
        inplace=True
    )
```

Here, we can see that we are using less code.

3. Now, we can do our final process, which is merging with the vulnerabilities and then calculating the total probability with the following code:

```
model = pd.merge(
    model, vulnerabilities,
    how="inner", on="intensity_bin_id"
)
model.rename(
    columns={"probability": \
        "vulnerability_probability"},
    inplace=True
)
model["total_prob"] = \
    model["footprint_probability"] * \
        model["vulnerability_probability"]
return model
```

With this, our Python model is now complete. We can now move on to our next step of building our random event ID generator functions.

Building a random event ID generator function

When it comes to our Rust implementation, we need a list of integers. For our Python model, we need to pass in a list of dictionaries with an event ID stored in it. We can define these functions with the following code:

```
def generate_event_ids_for_python(number_of_events):
    return [{"event_id": random.randint(1, 4)} for _
        in range(0, number_of_events)]


def generate_event_ids_for_rust(number_of_events):
    return [random.randint(1, 4) for _
        in range(0, number_of_events)]
```

Now that we have everything we need, we can carry out the final step of testing our implementations.

Timing our Python and Rust implementations with a series of different data sizes

We now have everything we need to test our Rust and Python implementation. Running both Python and Rust models with timing can be done by carrying out the following steps:

1. To test our implementation, we define our entry point and all the data structures for our time graph with the following code:

```
if __name__ == "__main__":
    x = []
    python_y = []
    rust_y = []
```

2. For our testing data, we are going to loop through a list of integers from 10 to 3000 in steps of 10 with the following code:

```
for i in range(10, 3000, 10):
    x.append(i)
```

Both Python and Rust implementations will be running the same event ID dataset sizes, which is why we only have one x vector. We can now test our Python implementation with the following code:

```
python_event_ids = \
    generate_event_ids_for_python(
        number_of_events=i
    )
python_start = time.time()
python_construct_model(event_ids= \
    python_event_ids)
python_finish = time.time()
python_y.append(python_finish - python_start)
```

Here, we generate our ID dataset to the size of the integer of the loop. We then start our timer, construct our model in Python, finish the timer, and add the time taken to our Python data list.

3.   We take the same approach with our Rust test with the following code:

```
rust_event_ids = generate_event_ids_for_rust(
    number_of_events=i
)
rust_start = time.time()
construct_model(rust_event_ids)
rust_finish = time.time()
rust_y.append(rust_finish - rust_start)
```

Our data collection is now complete.

4.   All we need to do is plot the results when the loop has finished with the following code:

```
plt.plot(x, python_y)
plt.plot(x, rust_y)
plt.show()
```

We have now written all the code for testing, which should display a graph like this:

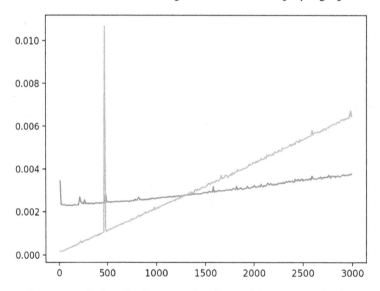

Figure 8.4 – Rust versus Python for the time taken for model generation for the size of data

In the preceding figure, we see that initially, our Rust implementation is faster than our Python pandas implementation. However, once we get past the 1,300 mark, our Rust model gets slower than our Python pandas model. This is because our code does not scale well. We are performing loops within loops. In our pandas model, we vectorize our total probability. pandas is a well-written module where multiple developers have optimized the merge functions.

Therefore, although our Rust code will be faster than Python and pandas code, if our implementation is sloppy and does not scale well, we may even be slowing down our program. I have seen poorly implemented C++ be beaten by Python pandas. Understanding this nuance is important when trying to implement Rust in your system. Rust is a new language, and colleagues will be let down if you promise big gains, poorly implement code, and result in slower performance after burning a lot of time coding implementation in Rust.

Seeing that this is a book about building Python packages in Rust as opposed to data processing in Rust, this is where we stop. However, Xavier Tao implemented an efficient merge process in Rust, resulting in Rust taking 75% less time and 78% less memory. This is noted in the *Further reading* section. There is also a Rust implementation of pandas called **Polars**, which also has Python bindings. It is faster than standard pandas, and this documentation is also listed in the *Further reading* section.

The takeaway message here is that Rust enables us to build fast memory-efficient solutions, but we must be careful with our implementation and test to see whether what we are doing is sensible. We should be careful, especially if we are trying to build a solution from scratch that has an optimized solution in an existing Python package.

Summary

In this chapter, we went through the basics of building a simple catastrophe model. We then broke down the logic and converted it into steps so that we could build the catastrophe model in Rust. This included taking in paths, loading data from files, including data in our package, and building a Python interface so that our users do not have to know about what is going on under the hood when constructing a model. After all of this, we tested our module and ensured that we kept increasing the data size of the test to see how it scales. We saw that, initially, our Rust solution was faster because Rust is faster than Python and pandas. However, our implementation did not scale well, as we did a loop within a loop for our merge.

As the data size increased, our Rust code ended up being slower. In previous chapters, we have shown multiple times that Rust implementations are generally faster. However, this does not counteract the effects of bad code implementation. If you are relying on a Python third-party module to perform a complex process, it probably is not a good idea to rewrite it in Rust for performance gains. If a Rust crate is not available for the same solution, then it is probably best to leave that part of the solution to the Python module.

In the next chapter, we will be building a Flask web application to lay the groundwork for applying Rust to a Python web application.

Further reading

- Polars documentation for Rust Crate Polars (2021): `https://docs.rs/polars/0.15.1/polars/frame/struct.DataFrame.html`

- *Data Manipulation: Pandas vs Rust, Xavier Tao* (2021): `https://able.bio/haixuanTao/data-manipulation-pandas-vs-rust--1d70e7fc`

Section 3: Infusing Rust into a Web Application

At this point, all the key areas have been covered in terms of using Rust practically in Python code. In this section, we will apply all of what we have learned thus far in a practical project. We achieve this by injecting our Python packages written in Rust into every aspect of a web application that can be deployed in Docker.

This section comprises the following chapters:

- *Chapter 9, Structuring a Python Flask App for Rust*
- *Chapter 10, Injecting Rust into a Python Flask App*
- *Chapter 11, Best Practices for Integrating Rust*

9

Structuring a Python Flask App for Rust

In the previous chapter, we managed to solve a real-world problem with Rust. However, we also learned an important lesson, that is, the good implementation of code, such as adding vectors or merging dataframes, along with third-party modules, such as NumPy, can outperform badly implemented self-coded Rust solutions. However, we know that comparing implementation to implementation, Rust is a lot faster than Python. We already understand how to fuse Rust with a standard Python script. However, Python is used for more than just running scripts. A popular use for Python is in web applications.

In this chapter, we will build a Flask web application with NGINX, a database, and a message bus implemented by the Celery package. This message bus will allow our application to process heavy tasks in the background while we return a web HTTP request. The web application and message bus will be wrapped in Docker containers and deployed to docker-compose. However, nothing is preventing us from deploying the application onto a cloud platform if desired.

In this chapter, we will cover the following topics:

- Building a basic Flask application
- Defining a database access layer
- Building a message bus

This chapter will enable us to build a foundation for deployable Python web applications that have a range of features and services. This foundation allows us to discover how to fuse Rust with Python web applications that are wrapped in Docker containers.

Technical requirements

The code and data for this chapter can be found at `https://github.com/ PacktPublishing/Speed-up-your-Python-with-Rust/tree/main/ chapter_nine`.

In addition to this, we will be using `docker-compose` on top of Docker to orchestrate our Docker containers. This can be installed by following the instructions at `https:// docs.docker.com/compose/install/`.

In this chapter, we will be building a Docker-contained Flask application, which is available via the GitHub repository at `https://github.com/maxwellflitton/ fib-flask`.

Building a basic Flask application

Before we begin adding any additional features such as a database to an application, we have to ensure that that we can get a basic Flask application up and running with everything that we need. This application will take in a number and return a Fibonacci number. Additionally, we will need to make sure that this application can run in its own Docker container if we were to deploy it. By the end of this section, our application should have the following structure:

```
├── deployment
│   ├── docker-compose.yml
│   └── nginx
│       ├── Dockerfile
│       └── nginx.conf
├── src
│   ├── Dockerfile
│   ├── __init__.py
│   ├── app.py
│   ├── fib_calcs
│   │   ├── __init__.py
│   │   └── fib_calculation.py
│   └── requirements.txt
```

Here, you can see that the application is housed in the `src` directory. When running our application, we must ensure that the `PYTHONPATH` path is set to `src`. The code required for our deployment exists in the `deployment` directory. To build an application so that it can run in Docker, perform the following steps:

1. Build an entry point for our application.
2. Build a Fibonacci number calculation module.
3. Build a Docker image for our application.
4. Build our NGINX service.

Once we have completed all of these steps, we will have a basic Flask application that can be run on a server. Now, let's explore each of these steps in detail in the following subsections.

Building an entry point for our application

Here are the steps to perform:

1. Before we can build our entry point, we need to install the Flask module using the following command:

    ```
    pip install flask
    ```

2. Once this is done, we have all we need to create a basic Flask application by defining the entry point in the `src/app.py` file using the following code:

    ```python
    from flask import Flask

    app = Flask(__name__)

    @app.route("/")
    def home():
        return "home for the fib calculator"

    if __name__ == "__main__":
        app.run(use_reloader=True, port=5002, \
            threaded=True)
    ```

Here, observe that we can define a basic route with the decorator. We can run our application by running the `src/app.py` script; this will run our server locally, enabling us to access all of the routes that we have defined. Passing the `http://127.0.0.1:5002` URL into our browser will give us the following view:

home for the fib calculator

Figure 9.1 – The main view of our local Flask server

Now that our basic server is running, we can move on to build a Fibonacci number calculator module.

Building our Fibonacci number calculator module

Here are the steps to perform:

1. In this example, our application is simple. As a result, we can define our module's functionality within one file in one class. We define it within the `src/fib_calcs/fib_calculation.py` file using the following code:

```python
class FibCalculation:

    def __init__(self, input_number: int) -> None:
        self.input_number: int = input_number
        self.fib_number: int = self.recur_fib(
            n=self.input_number
        )
    @staticmethod
    def recur_fib(n: int) -> int:
        if n <= 1:
            return n
        else:
            return (FibCalculation.recur_fib(n - 1) +
                    FibCalculation.recur_fib(n - 2))
```

Here, notice that our class merely takes in an input number and automatically populates the self.fib_number attribute with the calculated Fibonacci number.

2. Once that is done, we can define a view that accepts an integer through the URL, passes it to our FibCalculation class, and returns the calculated Fibonacci number as a string to the user in our src/app.py file using the following code:

```
from fib_calcs.fib_calculation import FibCalculation

. . .

@app.route("/calculate/<int:number>")
def calculate(number):
    calc = FibCalculation(input_number=number)
    return f"you entered {calc.input_number} " \
           f"which has a Fibonacci number of " \
           f"{calc.fib_number}"
```

3. Rerunning our server and passing the http://127.0.0.1:5002/calculate/10 URL into our browser will give us the following view:

you entered 10 which has a Fibonacci number of 55

Figure 9.2 – Calculating the view of our local Flask server

Now our application performs its intended purpose: it calculates a Fibonacci number based on the input. There is more to views with Flask; however, this book is not a web development textbook. If you want to learn how to build more comprehensive API endpoints, we advise that you look into the Flask API and Marshmallow packages. References to both are available in the *Further reading* section. Now, we need to make our application deployable so that we can use it in a range of settings for our next step.

Building a Docker image for our application

For our application to be usable, we must build a Docker image of our application that accepts requests. Then, we must protect it with another container call that acts as an ingress. NGINX performs load balancing, caching, streaming, and the redirecting of traffic. Our application will be run using the Gunicorn package, which, essentially, runs multiple workers of our application at the same time. For each request, NGINX asks which Gunicorn worker the request should go to and redirects it, as shown in the following diagram:

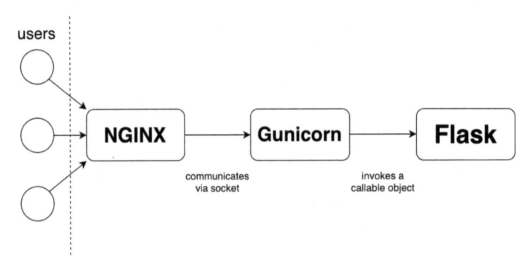

Figure 9.3 – The flow of requests for our application

We can achieve the layout defined in the preceding diagram by performing the following steps:

1. Before we build a Docker image, we must make sure the requirements for our application are handled. Therefore, we must install Gunicorn using `pip` with the following command:

    ```
    pip install gunicorn
    ```

2. We must make sure that we are in the `src` directory because we are going to dump all of our application dependencies into a file called `requirements.txt` using the following command:

    ```
    pip freeze > requirements.txt
    ```

    This gives us a text file with a list of all the dependencies that our application needs in order to run. Right now, all we need is Flask and Gunicorn.

3. With this, we can start coding our Docker file so that we can build application images of our application. First, in our `src/Dockerfile` file, we should define the operating system that is required with the following code:

```
FROM python:3.6.13-stretch
```

This means that our image is running a stripped-down version of Linux with Python installed.

4. Now that we have the correct operating system, we should define our app's directory and copy all of our application files into the image using the following code:

```
# Set the working directory to /app
WORKDIR /app

# Copy the current directory contents into the
  container at /app
ADD . /app
```

5. Now that all of our application files are in the image, we install system updates and then install the `python-dev` package. This is so that we can include extensions with the code given here:

```
RUN apt-get update -y
RUN apt-get install -y python3-dev python-dev gcc
```

This will enable us to compile our Rust code in our application and use database binaries.

Our system has now been set up, so we can move on to install our requirements using the following code:

```
RUN pip install --upgrade pip setuptools wheel
RUN pip install -r requirements.txt
```

Everything is in place to define our system. Nothing is stopping us from running our application.

6. To do this, we expose the port and run our application using the following code:

```
EXPOSE 5002

CMD ["gunicorn", "-w 4", "-b", "0.0.0.0:5002", \
  "app:app"]
```

Note that when we create a container from the image, we run CMD with the parameters defined in the list. We state that we have four workers with the -w 4 parameter. Then, we define the URL and port that we are listening to. Our final parameter is app:app. This states that our application is housed in the app.py file, and our application in that file is the Flask object under the variable name of app.

7. We can now build our application image using the following command:

```
docker build . -t flask-fib
```

The result is a very long stream of console printouts, but essentially, our Docker image is built with the flask-fib tag.

8. We can then inspect our images using the following command:

```
docker image ls
```

Running this command gives us an image that has been created in the following form:

REPOSITORY	TAG	IMAGE ID
flask-fib	latest	0cdb0c979ac1
CREATED	SIZE	
33 minutes ago	1.05GB	

This is important. We will need to reference our image later when we are running our application with NGINX, which we will define next.

Building our NGINX service

When it comes to Docker and NGINX, we are lucky in that we do not have to build a Dockerfile that defines our NGINX image. NGINX has released an official image that we can download and use for free. However, we do have to alter its configuration. NGINX is fairly important; this is because it gives us the ability to control how incoming requests are processed. We can redirect the requests to different services depending on parts of the URL. Additionally, we can control the size of the data, the duration of the connection, and configure HTTPS traffic. NGINX can also act as a load balancer. In this example, we are going to configure NGINX in the simplest format to get it running. However, it must be noted that NGINX is a vast topic in itself; a reference to a useful NGINX book is provided in the *Further reading* section.

We can build our NGINX service and connect it to our Flask application by performing the following steps:

1. We will configure our NGINX container with what we code in the `deployment/`
    `nginx/nginx.conf` file. In this file, we declare our worker processes and error
    log, as follows:

    ```
    worker_processes  auto;
    error_log  /var/log/nginx/error.log warn;
    ```

    Here, we have defined `worker_processes` as `auto`. This is where we
    automatically detect the number of CPU cores available, setting the number of
    processes to the number of CPU cores.

2. Now, we have to define the maximum number of connections that a worker can
    entertain at a time using the following code:

    ```
    events {
        worker_connections  512;
    }
    ```

    It must be noted that the number that is chosen here is the default number
    for NGINX.

3. All that is now left for us to do is to define our HTTP listener. This can be achieved
    with the following code:

    ```
    http {
        server {
            listen 80;

            location / {
                proxy_pass http://flask_app:5002/;
            }
        }
    }
    ```

Here, observe that we listen to port 80, which is the standard outside listening port. Then, we state that if there is any pattern to our URL, we pass it to our `flask_app` container at port 5002. We can stack multiple locations in the `http` section if we wish. For instance, if we have another app, we can route the request to the other application if the URL tail starts with `/another_app/` using the following code:

```
location /another_app {
        proxy_pass http://another_app:5002/;
}
location / {
        proxy_pass http://flask_app:5002/;
}
```

Our configuration file for our NGINX is complete. Again, there are many more configuration parameters; we are just running the bare minimum. More resources on these parameters are signposted in the *Further reading* section. Considering that our NGINX configuration file is complete, for the next step, we have to run it alongside our Flask application.

Connecting and running our Nginx service

To run our application and NGINX together, we will be using `docker-compose`. This allows us to define multiple Docker containers at the same time that can talk to each other. Nothing is stopping us from running `docker-compose` on a server to achieve a basic setup. However, more advanced systems such as Kubernetes can help with the orchestration of Docker containers across multiple servers if needed. In addition to this, different cloud platforms offer out-of-the-box load balancers. Perform the following steps:

1. In our `deployment/docker-compose.yml` file, we state what version of `docker-compose` we are using with the following code:

    ```
    version: "3.7"
    ```

2. Now that this has been implemented, we can define our services along with our first service, which is our Flask application. This is defined with the following code:

    ```
    services:

        flask_app:
            container_name: fib-calculator
            image: "flask-fib:latest"
            restart: always
    ```

```
        ports:
            - "5002:5002"
        expose:
            - 5002
```

In the preceding code, we reference the image that we built with the latest release. For instance, if we changed the image and rebuilt it, then our `docker-compose` setup would use this. We also give it a container name, so we know the container status when checking the running containers. Additionally, we state that we accept traffic through port `5002`, and we route it to our container's port `5002`. Because we have chosen this path, we also expose port `5002`. If we run our `docker-compose` setup now, we could access our application with the `http://localhost:5002` URL. However, if this was running on a server and port `5002` was not accessible to outside traffic, then we would not be able to access it.

3. Considering this, we can define our NGINX in our `deployment/docker-compose.yml` file using the following code:

```
    nginx:
        container_name: 'nginx'
        image: "nginx:1.13.5"
        ports:
            - "80:80"
        links:
            - flask_app
        depends_on:
            - flask_app
        volumes:
            - ./nginx/nginx.conf:/etc/nginx/nginx.conf
```

Here, you can see that we rely on the third-party NGINX image and that we route the outside port of 80 to port 80. Also, we link to our Flask application, and we depend on it, meaning that `docker-compose` will ensure that our Flask application is up and running before we run our NGINX service. In the `volumes` section, we replace the standard configuration file with the configuration file that we defined in the previous step. As a result, our NGINX service will run the configuration that we defined. It must be noted that this configuration switch will happen every time we run `docker-compose`. This means that if we change our configuration file and then run `docker-compose` again, we will see the changes. So, we have done everything to get our application up and running. Now we can test it.

4. Testing our application is as easy as running the following command:

```
Docker-compose up
```

Our services will boot up, and we will get the following printout:

```
Starting fib-calculator ... done
Starting nginx          ... done
Attaching to fib-calculator, nginx
fib-calculator | [2021-08-20 18:43:14 +0000] [1]
[INFO]
Starting gunicorn 20.1.0
fib-calculator | [2021-08-20 18:43:14 +0000] [1]
[INFO]
Listening at: http://0.0.0.0:5002 (1)
fib-calculator | [2021-08-20 18:43:14 +0000] [1]
[INFO]
Using worker: sync
fib-calculator | [2021-08-20 18:43:14 +0000] [8]
[INFO]
Booting worker with pid: 8
fib-calculator | [2021-08-20 18:43:14 +0000] [9]
[INFO]
Booting worker with pid: 9
. . .
nginx          | /docker-entrypoint.sh: Configuration
complete;
ready for start up
```

Here, notice that both of our services spin up without any problems. Our Flask application starts Gunicorn, starts listening at port 5002, and boots up workers to process requests. Following this, our NGINX service looks for a range of configurations before concluding that the configuration is complete and that it is ready to start up. Also, note that the NGINX started after our Flask application was started. This is because we stated that our NGINX was dependent on our Flask application when building our docker-compose file.

Now, we can directly hit our localhost URL without having to specify a port because we are listening to the outside port of 80 with our NGINX. This gives us results similar to the following:

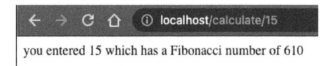

you entered 15 which has a Fibonacci number of 610

Figure 9.4 – Interacting with our fully containerized Flask application

Now we have a fully containerized application that runs. This is at a ready state, so in the next chapter, we can test to see whether our Rust integration with our application will actually work in a real-life scenario. Now that we have gotten our application running, we can move on to build our data access layer. This will allow us to store and get data from a database.

Defining our data access layer

Now we have an application that takes in a number and calculates a Fibonacci number based on it. However, a database lookup is quicker than a calculation. We will use this fact to optimize our application by initially performing a database lookup when a number is submitted. If it is not there, we calculate the number, store it in the database, and then return it to the user. Before we start building, we will have to install the following packages using pip:

- pyml: This package helps in loading parameters for our application from a .yml file.
- sqlalchemy: This package enables our application to map Python objects to databases for storing and querying.
- alembic: This package helps in tracking and applying changes to the database from the application.
- psycopg2-binary: This is the binary that will enable our application to connect to the database.

Now that we have installed all that we need, we can enable our application to store and get Fibonacci numbers by performing the following steps:

1. Define a PostgreSQL database in `docker-compose`.

2. Build a config loading system.

3. Define a data access layer.

4. Build database models.

5. Set up the application database migration system.

6. Apply the database access layer to the fib calculation view.

Once we have completed these steps, our application will take the following form:

```
├── deployment
|     . . .
├── docker-compose.yml
├── src
|     . . .
|     ├── config.py
|     ├── config.yml
|     ├── data_access.py
|     ├── fib_calcs
|     |     . . .
|     ├── models
|     |     ├── __init__.py
|     |     └── database
|     |           ├── __init__.py
|     |           └── fib_entry.py
|     └── requirements.txt
```

Our deployment file structure has not changed. We have added a `docker-compose.yml` file to our root as it will enable us to access the database when we are developing our application. In addition to this, we have added a data access file to enable us to connect to the database along with a `models` module to enable mapping objects to the database. This structure will result in a containerized Flask application that has access to a database. Next, we will begin defining our Docker container for our database.

Defining a PostgreSQL database in docker-compose

To define our database container, we apply the following code to both the `deployment/docker-compose.yml` file and the `docker-compose.yml` file:

```yaml
Postgres:
    container_name: 'fib-dev-Postgres
    image: 'postgres:11.2'
    restart: always
    ports:
        - '5432:5432'
    environment:
        - 'POSTGRES_USER=user'
        - 'POSTGRES_DB=fib'
        - 'POSTGRES_PASSWORD=password'
```

Here, you can observe that we are relying on the official third-party Postgres image. Instead of defining a configuration file, as we did with the NGINX service, we define the password, database name, and user using the environment variables. When we are running our local environment and developing our application, we will run our `docker-compose` file in the root. Now we have defined our database; in the next section, we can build our config system.

Building a config loading system

Essentially, our configuration system loads parameters from a `.yml` file inside the Flask application by performing these steps:

1. Our application might require different parameters depending on the system. Because of this, we must build an object that loads parameters from a `.yml` file and serves them as a dictionary throughout the application. In our `src/config.py` file, first, we import what we need with the following code:

```python
import os
import sys
from typing import Dict, List


import yaml
```

We will be using the `sys` module to take in the arguments that were passed into our application while running it. We use the `os` module to check whether the config file that we have specified in the arguments exists.

2. Our global parameters object can be built using the following code:

```python
class GlobalParams(dict):

    def __init__(self) -> None:
        super().__init__()
        self.update(self.get_yml_file())

    @staticmethod
    def get_yml_file() -> Dict:
        file_name = sys.argv[-1]
        if ".yml" not in file_name:
            file_name = "config.yml"

        if os.path.isfile(file_name):
            with open("./{}".format(file_name)) as \
                file:
                    data = yaml.load(file,
                            Loader=yaml.FullLoader)
            return data
        raise FileNotFoundError(
            "{} config file is not available".
                format(file_name)
        )
    @property
    def database_meta(self) -> Dict[str, str]:
        db_string: str = self.get("DB_URL")
        buffer: List[str] = db_string.split("/")
        second_buffer: List[str] = buffer[- \
            2].split(":")
        third_buffer: List[str] = \
            second_buffer[1].split("@")
        return {
            "DB_URL": db_string,
```

```
        "DB_NAME": buffer[-1],
        "DB_USER": second_buffer[0],
        "DB_PASSWORD": third_buffer[0],
        "DB_LOCATION":f"{third_buffer[1]} \
         :{second_buffer[-1]}",
    }
```

Here, you can observe that our `GlobalParams` class directly inherits from the dictionary class. This means that we have all the functionality of a dictionary. In addition to this, note that we do not pass any arguments into our Python program specifying which `.yml` file to load; instead, we simply revert to the standard `config.yml` file. This is because we will use our configuration file for migrations to the database. It will be difficult to pass in our parameters when performing database migrations. If we want to change the configuration, it is best to get the new data and write it to the config file.

3. Now that our config parameters class has been defined, we can add the database URL to our `src/config.yml` file using the following code:

```
DB_URL: \
"postgresql://user:password@localhost:5432/fib"
```

Now that we have access to our database URL, in the next step, we can build our database access layer.

Building our data access layer

Our database access will be defined in the `src/data_access.py` file. Once this is done, we can import the data access layer from the `src/data_access.py` file anywhere in the Flask application. This is so that we can access the database anywhere inside the Flask application. We can build this by performing the following steps:

1. First of all, we have to import what we need using the following code:

```
from flask import _app_ctx_stack
from sqlalchemy import create_engine
from sqlalchemy.ext.declarative import
declarative_base
from sqlalchemy.orm import sessionmaker,
scoped_session
from config import GlobalParams
```

Here, we will use the `_app_ctx_stack` object to ensure that our session is in the context of the Flask request. Following this, we import all of the other `sqlalchemy` dependencies to ensure that our access has a session maker and an engine. We have to avoid going into excessive detail with database management as this book focuses on fusing Rust with Python and we are merely using SQLAlchemy to explore database integration with Rust. However, we should be able to get a feel for what the session, engine, and base do.

2. Now that we have imported everything we need, we can build our database engine using the following code:

```python
class DbEngine:

    def __init__(self) -> None:
        params = GlobalParams()
        self.base = declarative_base()
        self.engine = create_engine(params.get
            ("DB_URL"),
                                    echo=True,
                                    pool_recycle=3600,
                                    pool_size=2,
                                    max_overflow=1,
                                    connect_args={
                                    'connect_timeout': 5
                                    })
        self.session = scoped_session(sessionmaker(
            bind=self.engine
        ), scopefunc=_app_ctx_stack)
        self.url = params.get("DB_URL")


dal = DbEngine()
```

Now we have a class that can give us database sessions, a database connection, and a base. However, it must be noted that we initiated the DbEngine class and assigned it to the dal variable; however, we didn't import the DbEngine class outside of this file. Instead, we import the dal variable to be used for interactions with the database. If we import the DbEngine class outside of this file during initiation and use it whenever we want to interact with the database, we will create multiple database sessions per request and these sessions will struggle to close. Even something as small as a couple of users will grind your database to a halt with too many hanging connections. Now that our database connection has been defined, in the next step, we can move on to build our database models.

3. In our database model, we can have a unique ID, input number, and fib number. Our model is defined in the src/models/database/fib_entry.py file with the following code:

```python
from typing import Dict
from sqlalchemy import Column, Integer
from data_access import dal


class FibEntry(dal.base):

    __tablename__ = "fib_entries"


    id = Column(Integer, primary_key=True)
    input_number = Column(Integer)
    calculated_number = Column(Integer)


    @property
    def package(self) -> Dict[str, int]:
        return {
            "input_number": self.input_number,
            "calculated_number": \
                self.calculated_number
        }
```

Here, you can see that the code is straightforward. We pass `dal.base` through our model to add the model to the metadata. Then, we define the table name that will be in the database and model fields, which are `id`, `input_number`, and `calculated_number`. Our database model has now been defined, so we can import and use it throughout our application. Additionally, we will use this in the next step to manage the database migrations.

Setting up the application database migration system

Migrations are a useful tool for keeping track of all the changes made to our database. If we make a change in a database model or define one, we need to translate those changes to our database. We can achieve this by performing the following steps:

1. For our database management, we are going to lean on the `alembic` package. Once we have navigated inside the `src/` directory, we run the following command:

   ```
   alembic init alembic
   ```

   This will generate a range of scripts and files. We are interested in the `src/alembic/env.py` file; we are going to alter this so that we can connect our `alembic` scripts and commands to our database.

2. Next, we must import the `os` and `sys` modules, as we will be using them to import our models and load our configuration file. We import the modules using the following code:

   ```
   import sys
   import os
   ```

3. Following this, we use the `os` module to append the path that is in the `src/` directory with the following code:

   ```
   from alembic import context


   # this is the Alembic Config object, which provides
   # access to the values within the .ini file in use.
   Config = context.config


   # Interpret the config file for Python logging.
   # This line sets up loggers basically.
   fileConfig(config.config_file_name)
   # add the src to our import path
   ```

```
sys.path.append(os.path.join(
    os.path.dirname(os.path.abspath(__file__)),
    "../")
)
```

4. Now that we have configured our import path, we can import our parameters and database engine. Then, we add our database URL to our `alembic` database URL using the following code:

```
# config the database url for migrations
from config import GlobalParams
params = GlobalParams()
section = config.config_ini_section
db_params = params.database_meta


config.set_section_option(section, 'sqlalchemy.url',
                            params.get('DB_URL'))


from data_access import dal
db_engine = dal


from models.database.fib_entry import FibEntry


target_metadata = db_engine.base.metadata
```

5. With this, you can observe that the autogenerated function gets our config, which then executes the migrations with the following code:

```
def run_migrations_offline():
    url = config.get_main_option("sqlalchemy.url")
    context.configure(
        url=url,
        target_metadata=target_metadata,
        literal_binds=True,
        dialect_opts={"paramstyle": "named"},
        render_as_batch=True
    )
```

```
with context.begin_transaction():
    context.run_migrations()
```

6. Now that we have our configuration system linked up to our database migrations, we have to make sure `docker-compose` is running because our database has to be live. We can generate a migration using the following command:

```
alembic revision --autogenerate -m create-fib-entry
```

This gives us the following printout:

```
INFO  [alembic.runtime.migration] Context impl
PostgresqlImpl.
INFO  [alembic.runtime.migration] Will assume
transactional DDL.
INFO  [alembic.autogenerate.compare] Detected added
table 'fib_entries'
```

You can observe that in our `src/alembic/versions/` file, there is an autogenerated script that creates our table with the following code:

```
# revision identifiers, used by Alembic.
Revision = '40b83d85c278'
down_revision = None
branch_labels = None
depends_on = None


def upgrade():
    op.create_table('fib_entries',
    sa.Column('id', sa.Integer(), nullable=False),
    sa.Column('input_number', sa.Integer(),\
      nullable=True),
    sa.Column('calculated_number', sa.Integer(), \
      nullable=True),
    sa.PrimaryKeyConstraint('id')
    )
def downgrade():
    op.drop_table('fib_entries')
```

Here, if we upgrade, the `upgrade` function will run, and if we downgrade, the `downgrade` function will run. We can upgrade our database using the following command:

```
alembic upgrade head
```

This gives us the following printout:

```
INFO   [alembic.runtime.migration] Context impl
PostgresqlImpl.
INFO   [alembic.runtime.migration] Will assume
transactional DDL.
INFO   [alembic.runtime.migration] Running upgrade   ->
40b83d85c278, create-fib-entry
```

Our migration has worked. In the next step, we will interact with the database in our application.

Building database models

Now that we have a database that has our application models applied to it, we can interact with our database in the application. This can be done by importing our data access layer and data model into the view that is using them and, well, use them:

1. For our example, we will be implementing our view inside the `src/app/app.py` file. First, we import the data access layer and model using the following code:

```
from data_access import dal
from models.database.fib_entry import FibEntry
```

With these imports, we can alter our calculation view to check whether the number exists in the database and return the number from the database if it does.

2. If it is not available in the database, then we calculate it, save the result in the database, and return the result using the following code:

```
@app.route("/calculate/<int:number>")
def calculate(number):
    fib_calc = dal.session.query(FibEntry).filter_by(
            input_number=number).one_or_none()
    if fib_calc is None:
        calc = FibCalculation(input_number=number)
        new_calc = FibEntry(input_number=number,
```

```
                    calculated_number=calc.fib_number)
        dal.session.add(new_calc)
        dal.session.commit()


        return f"you entered {calc.input_number} " \
               f"which has a Fibonacci number of " \
               f"{calc.fib_number}"
    return f"you entered {fib_calc.input_number} " \
           f"which has an existing Fibonacci number of " \
           f"{fib_calc.calculated_number}"
```

Here, you can observe that our interactions with the database are straightforward.

3. Now we have to make sure that when our request has finished, our database sessions
    are expired, closed, and removed using the following code:

```
@app.teardown_request
def teardown_request(*args, **kwargs):
    dal.session.expire_all()
    dal.session.remove()
    dal.session.close()
```

So, we have a safe and fully functioning interaction with our database. You are now aware
of the fundamentals of interacting with a database using our application. You can achieve
other, more complex database queries by reading the SQLAlchemy documentation about
the specifics of the database, other database queries, and insertions as a way to map
syntax. If we run our application locally and hit our calculation view twice, we will get the
first and second results, as shown in the following screenshot:

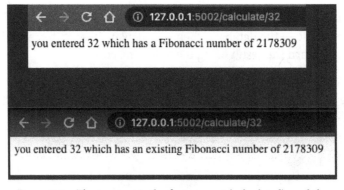

Figure 9.5 – The top part is the first request (calculated), and the
bottom part is the second request (database call)

Our database is working the way we expect it to. Now the application is fully functioning, and you can move on to the next section if you wish, as this is enough to test Rust code in a Flask application, which we will do in the next chapter. However, if you want to understand how we apply the database in our deployment section, we will cover this next.

Applying the database access layer to the fib calculation view

Adding a database to our deployment is a matter of adding it to our `docker-compose` deployment and updating our configuration file to map to the database service in the `docker-compose` deployment. We can achieve this by performing the following steps:

1. First, we have to refactor our `deployment/docker-compose.yml` file using the following code:

```
services:


    flask_app:
        container_name: fib-calculator
        image: "flask-fib:latest"
        restart: always
        ports:
          - "5002:5002"
        expose:
            - 5002
        depends_on:
          - postgres
        links:
            - postgres


    nginx:
        . . .


    postgres:
        container_name: 'fib-live-postgres'
        image: 'postgres:11.2'
        restart: always
        ports:
```

```
        - '5432:5432'
    environment:
        - 'POSTGRES_USER=user'
        - 'POSTGRES_DB=fib'
        - 'POSTGRES_PASSWORD=password'
```

You can observe that we have a slightly different name for our database container. This is to ensure that there are no clashes with our development database. Additionally, we have declared that our Flask application is dependent on and linked to our database.

2. We also have to point our Flask application to the Docker database. To do this, we have to have a different configuration file for our Flask application. We can manage the configuration file switch in `src/Dockerfile` for the Flask application. This can be done using the following code:

```
# Copy the current directory contents into the
    container at /app
ADD . /app

RUN rm ./config.yml
RUN mv live_config.yml config.yml
. . .
```

Here, we remove the `config.yml` file and then change the filename of the `live_config.yml` file to `config.yml`.

3. Then, we have to make our `src/live_config.yml` file with the following content:

```
DB_URL: "postgresql://user:password@postgres:5432/fib"
```

Here, we have changed `@localhost` to `@postgres` because the classification of our service is called `postgres`.

4. Following this, we can rebuild our Flask image using the following command:

```
docker build . -t flask-fib
```

5. Now we can run our docker-compose deployment, but we will have to run our migrations while our docker-compose deployment is running. This is because our Flask application will not cause an error if it is out of sync with the database until we try and make a query to the database, so running docker-compose before migrating is fine if we make no requests. When we make the migration, we must do this while the database in docker-compose is running; otherwise, the migration will not be able to connect to the database. We can run the migration while docker-compose is running using the following command:

```
docker exec -it fib-calculator alembic upgrade head
```

This runs the migrations on our Flask container. It is not advised that you only have your live configuration files within your application code. A favorite method of mine is to upload an encrypted configuration file in AWS S3 and pull it in Kubernetes as it starts up a pod. This is beyond the scope of this book, as this is not a web development book. However, it is important to keep methods such as this in mind for further reading if needed.

Right now, there is not much to complain about when it comes to calculating Fibonacci numbers with our Flask application. However, when we try and calculate a large number, we will be waiting for a long time, and this will keep the request hanging. To prevent this from occurring, in the next section, we are going to implement a message bus. This is so that while our application is processing large numbers in the background, we can return a message telling the users to be patient.

Building a message bus

For this section, we will be using the Celery and Redis packages to build and run our message bus. Once we have completed this section, our mechanism will take a form that is similar to the following:

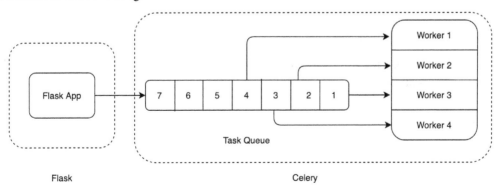

Figure 9.6 – A message bus with Flask and Celery

As shown in the preceding diagram, we have two processes running. One is running our Flask application, while the other is running `Celery`, which handles queuing and processing tasks. To make this work, we are going to perform the following steps:

1. Build a `Celery` broker for Flask.
2. Build a Fibonacci calculation task for **Celery**.
3. Update our calculation view with `Celery`.
4. Define our `Celery` service in Docker.

Before we embark on these steps, we have to install the following packages using `pip`:

- `Celery`: This is the message bus broker that we are going to use.
- `Redis`: This is the storage system that `Celery` is going to use.

Now that we have installed the requirements, we have to remember to update the `src/requirements.txt` file with `Celery` and Redis for our Docker builds. Now that we have all of our dependencies installed, we can start building our `Celery` broker, as demonstrated next.

Building a Celery broker for Flask

Essentially, our `Celery` broker is a storage system that will store data concerning the tasks we have sent to it. We can set up our storage system and connect it to our `Celery` system using the following steps:

1. We are going to build our own module when building our task queue. Inside the `src/` directory, our task queue module will take the following structure:

```
└── task_queue
    ├── __init__.py
    ├── engine.py
    └── fib_calc_task.py
```

Our `engine.py` file will host a constructor for `Celery` that considers the context of the Flask application.

2. We will build our Fibonacci calculation `Celery` task in the `fib_calc_task.py` file. In our `engine.py` file, we can build our constructor using the following code:

```python
from celery import Celery
from config import GlobalParams


def make_celery(flask_app):
    params = GlobalParams()
    celery = Celery(
        backend=params.get("QUEUE_BACKEND"),
        broker=params.get("QUEUE_BROKER")
    )
    celery.conf.update(flask_app.config)

    class ContextTask(celery.Task):
        def __call__(self, *args, **kwargs):
            with flask_app.app_context():
                return self.run(*args, **kwargs)

    celery.Task = ContextTask
    return celery
```

The `backend` and `broker` parameters will point to the storage; we will define them later. Here, you can observe that we must pass the Flask application into the function, construct the `Celery` class, and fuse a `Celery` task object with the Flask application context and then return it. When it comes to defining an entry point for running our `Celery` process, we should place it in the same file as our Flask application. This is because we want to use the same Docker build and, thus, image for the Flask application and `Celery` process.

3. To achieve this, we import our `Celery` constructor and pass the Flask application through it, in the `src/app.py` file, using the following code:

```python
. . .
from task_queue.engine import make_celery

app = Flask(__name__)
celery = make_celery(app)
. . .
```

4. Now, when we run our `Celery` broker, we will point it at our `src/app.py` file and the `Celery` object inside it. Additionally, we must define our backend storage system. Because we are using Redis, we can define these parameters in our `src/config.yml` file using the following code:

```
QUEUE_BACKEND: "redis://localhost:6379/0"
QUEUE_BROKER: "redis://localhost:6379/0"
```

Now that we have defined our `Celery` broker, in the next step, we can build our Fibonacci calculation task.

Building a Fibonacci calculation task for Celery

When it comes to running our `Celery` task, we need to build another constructor. However, instead of passing in our Flask application, we pass in our `Celery` broker. We can achieve this in the `src/task_queue/fib_calc_task.py` file using the following code:

```python
from data_access import dal
from fib_calcs.fib_calculation import FibCalculation
from models.database.fib_entry import FibEntry


def create_calculate_fib(input_celery):
    @input_celery.task()
    def calculate_fib(number):
        calculation = FibCalculation(input_number=number)
        fib_entry = FibEntry(
            input_number=calculation.input_number,
            calculated_number=calculation.fib_number
        )
        dal.session.add(fib_entry)
        dal.session.commit()
    return calculate_fib
```

The preceding logic is like our standard calculation view. We can import it into our `src/app.py` file and pass our `Celery` broker to it using the following code:

```python
. . .
from task_queue.engine import make_celery
```

```
app = Flask(__name__)
celery = make_celery(app)


from task_queue.fib_calc_task import create_calculate_fib
calculate_fib = create_calculate_fib(input_celery=celery)
. . .
```

Now that we have our task defined and fused with our `Celery` broker and Flask application, in the next step, we can add our `Celery` task to the calculation view if the number is too large.

Updating our calculation view

With our view, we must check to see whether our input number is less than 31 and not in the database. If it is, we run our standard existing code. However, if the input number is 30 or above, we will send the calculation to the `Celery` broker and return a message telling the user that it has been sent to the queue. We can do this using the following code:

```
@app.route("/calculate/<int:number>")
def calculate(number):
    fib_calc = dal.session.query(FibEntry).filter_by(
                        input_number=number).one_or_none()
    if fib_calc is None:
        if number < 31:
            calc = FibCalculation(input_number=number)
            new_calc = FibEntry(input_number=number,
                                calculated_number=calc.
                                fib_number)
            dal.session.add(new_calc)
            dal.session.commit()

            return f"you entered {calc.input_number} " \
                   f"which has a Fibonacci number of " \
                   f"{calc.fib_number}"
        calculate_fib.delay(number)
        return "calculate fib sent to queue because " \
               "it's above 30"
    return f"you entered {fib_calc.input_number} " \
```

```
            f"which has an existing Fibonacci number of " \
            f"{fib_calc.calculated_number}"
```

Now our `Celery` process with our task has been fully built. In the next step, we will define our Redis service in `docker-compose`.

Defining our Celery service in Docker

When it comes to our `Celery` service, remember that we used Redis as a storage mechanism. Considering this, we define our Redis service in our developed `docker-compose.yml` file using the following code:

```
. . .
    redis:
        container_name: 'main-dev-redis'
        image: 'redis:5.0.3'
        ports:
            - '6379:6379'
```

Now running our whole system in develop mode requires running our developed `docker-compose` file at the root of our project. Additionally, we run the Flask application by running our `app.py` file with Python, where `PYTHONPATH` is set to `src`.

Following this, we open another Terminal window, navigate the Terminal inside the `src` directory, and run the following command:

```
celery -A app.celery worker -l info
```

This is where we point `Celery` to the `app.py` file. We state that the object is called `Celery`, that it is a worker, and that the logging is at the `info` level. Running this gives us the following printout:

```
 -------------- celery@maxwells-MacBook-Pro.
--- ***** ----- local v5.1.2 (sun-harmonics)
-- ******* ---- Darwin-20.2.0-x86_64-i386-64bit
- *** --- * --- 2021-08-22 23:24:14
- ** ---------- [config]
- ** ---------- .> app:         __main__:0x7fd0796d0ed0
- ** ---------- .> transport:   redis://localhost:6379/0
- ** ---------- .> results:     redis://localhost:6379/0
- *** --- * --- .> concurrency: 4 (prefork)
```

```
-- ******* ---- .> task events: OFF (enable -E to
--- ***** -----    monitor tasks in this worker)
 -------------- [queues]
                .> celery  exchange=celery(direct)
 key=celery
 [tasks]
   . task_queue.fib_calc_task.calculate_fib
 [2021-08-22 23:24:14,385: INFO/MainProcess] Connected
 to redis://localhost:6379/0
 [2021-08-22 23:24:14,410: INFO/MainProcess] mingle:
 searching for neighbors
 [2021-08-22 23:24:15,476: INFO/MainProcess] mingle:
 all alone
 [2021-08-22 23:24:15,514: INFO/MainProcess]
 celery@maxwells-MacBook-Pro.local ready.
 [2021-08-22 23:24:39,822: INFO/MainProcess]
 Task task_queue.fib_calc_task.calculate_fib
 [c3241a5f-3208-48f7-9b0a-822c30aef94e] received
```

The preceding printout shows us that our task has been registered and that four processes have been spun up. Hitting the calculation view with our Celery processes using a number higher than 30 gives us the following view:

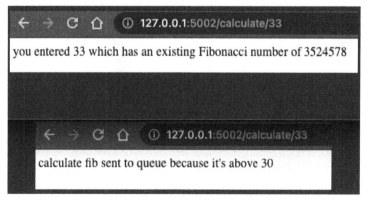

Figure 9.7 – The bottom shows the first request with Celery, and the top shows the second request with Celery

Our Flask application with a database and `Celery` message bus is now fully working locally. You can stop here if you wish, as this is enough to test Rust code in `Celery` in the next chapter. However, if you want to learn how to apply `Celery` to the deployment section, continue with this section.

Applying `Celery` to our `docker-compose` deployment is straightforward. Remember that we have the same entry point, so there is no need for a new image. Instead, all we have to do is change the command that we run when spinning up our `Celery` container. This can be done in our `deployment/docker-compose.yml` file using the following code:

```
. . .

  main_cache:
      container_name: 'main-live-redis'
      image: 'redis:5.0.3'
      ports:
          - '6379:6379'

  queue_worker:
      container_name: fib-worker
      image: "flask-fib:latest"
      restart: always
      entrypoint: "celery -A app.celery worker -l info"
      ports:
          - "5003:5003"
      expose:
          - 5003
      depends_on:
          - main_cache
      links:
          - main_cache
```

Here, you can observe that we pull the same image for our `queue_worker` service. However, we change the `CMD` tag in our Docker build using the `entrypoint` tag in `docker-compose`. So, when our `queue_worker` service is built, it will run the `Celery` command running the `Celery` workers, as opposed to running the Flask web application. Following this, we need to add some more parameters to our `live_config.yml` file using the following code:

```
QUEUE_BACKEND: "redis://main_cache:6379/0"
QUEUE_BROKER: "redis://main_cache:6379/0"
```

Here, we have named our Redis service as opposed to the localhost. This is so that our packaged `Celery` worker and Flask application will connect to our Redis service in the `docker-compose` deployment. After running the `docker-compose` deployment, we can repeat the requests demonstrated in *Figure 9.6* with `localhost` as opposed to `127.0.0.1:5002`. With this, our Flask application is ready to deploy with a database and task queue. Technically, our setup can be deployed and used on a server. I have done this, and it works just fine. However, for more advanced systems and control, it is advised that you carry out some further reading. Additional references about deploying Flask applications in Docker to cloud services such as Amazon Web Services are listed in the *Further reading* section.

Summary

In this chapter, we built a Python Flask application that had access to a database and message bus to allow the queuing of heavy tasks in the background. Following this, we wrapped our services in Docker containers and deployed them in a simple `docker-compose` file with NGINX. Additionally, we learned how to build our `Celery` worker and Flask application in the same Dockerfile using the same build. This made our code easier to maintain and deploy. We also managed our migrations for our database using `alembic` and a configuration file, which was then switched to another configuration file when we were deploying our application. While this is not a web development textbook, we have covered all of the essentials when it comes to structuring a Flask web application.

Further details regarding database queries, data serialization, or HTML and CSS rendering are covered, in a straightforward manner, in the Flask documentation. We have covered all of the difficult stuff. Now, we can experiment with Rust and how it can be fused with a Python web application, not just in a development setting but a live setting where the application is running in a Docker container while communicating with other Docker containers. In the next chapter, we will fuse Rust with our Flask application. This is so that it can work with the development and deployment settings.

Questions

1. What do we change in the URI when we switch from development to deployment on `docker-compose` to communicate with another service?

2. Why do we use configuration files?

3. Do we really need `alembic` to manage the database?

4. What do we have to do to our database engine to ensure our database does not get flooded with hanging sessions?

5. Do we need Redis for our `Celery` worker process?

Answers

1. We switch the `localhost` part of the URI to the tag of the `docker-compose` service.

2. Configuration files enable us to switch contexts easily; for instance, switching from development to live. Additionally, if our `Celery` service needs to talk to a different database for some reason, this can be done with minimal effort; simply changing the configuration file will work. It is also a security issue. Hardcoding database URIs will expose these credentials to anyone who has access to the code and will be in the GitHub repository history. Store the configuration file in a different space such as AWS S3, which gets pulled when the service is deployed.

3. Technically, no. We can simply write SQL scripts and run them in sequence. When I was working in financial technology, this was actually a thing that we had to do. While this can give you more freedom, it does take more time and is more error-prone. Using `alembic` will save you time, errors, and work for pretty much most of your needs.

4. We initiate our database engine once in the same file where our engine is defined. We never initiate it again, and we import this initiated engine anywhere we need. Not doing so will lead to our database to a grinding halt with dangling sessions and not very helpful error messages that will have you running around in circles on the internet with vague half-baked answers. Additionally, we have to close our sessions in the Flask teardown function for all requests.

5. Yes and no. We require a storage mechanism such as Redis; however, we can also use RabbitMQ or MongoDB instead of Redis if needed.

Further reading

- *Nginx HTTP Server – Fourth Edition: Harness the power of Nginx* by Fjordvald M. and Nedelcu C. (2018) (Packt)

- The official Flask documentation – Pallets (2021): `https://flask.palletsprojects.com/en/2.0.x/`

- *Hands-On Docker for Microservices with Python* by Jaime Buelta (2019) (Packt)

- *AWS Certified Developer – Associate Guide – Second Edition* by Vipul Tankariya and Bhavin Parmar (2019) (Packt)

- The SQLAlchemy query reference documentation (2021): `https://docs.sqlalchemy.org/en/14/orm/loading_objects.html`

10

Injecting Rust into a Python Flask App

In *Chapter 9, Structuring a Python Flask App for Rust*, we set up a basic Python web application in Flask that could be deployed using Docker. In this chapter, we are going to fuse Rust into every aspect of that web application. This means polishing our skills of defining Rust packages that can be installed using `pip`. With these packages, we are going to plug Rust code into our Flask and Celery containers. We are also going to directly interact with an existing database using Rust, without having to worry about migrations. This is because our Rust package is going to mirror the schema of the existing database. We will need a Rust `nightly` version to compile our package, so we will also learn how to manage Rust `nightly` when building our Flask image. We will also learn how to use Rust packages from private GitHub repositories.

In this chapter, we will cover the following topics:

- Fusing Rust into Flask and Celery
- Deploying Flask and Celery with Rust
- Deploying with a private GitHub repository
- Fusing Rust with data access
- Deploying Rust `nightly` in Flask

Learning about these topics will enable us to use our Rust packages in a Python web application so that it can be deployed in Docker. This will bring our Rust skills directly in line with the real world, enabling us to speed up Python web applications without having to rewrite our entire infrastructure. If you are a Python web developer, you will be able to turn up to work after reading this chapter and start injecting Rust into web applications to introduce fast, safe code without much risk.

Technical requirements

The following are the technical requirements for this chapter:

- The code and data for this chapter can be found at `https://github.com/ PacktPublishing/Speed-up-your-Python-with-Rust/tree/main/ chapter_ten`.

- In this chapter, you will be building a Docker-contained Flask application. This is available via the following GitHub repository: `https://github.com/ maxwellflitton/fib-flask`.

Fusing Rust into Flask and Celery

We will fuse Rust into our Flask application by installing our Rust Fibonacci calculation library using `pip`. We will then use it in our views and Celery tasks. This will speed up our Flask application without us having to make big changes to our infrastructure. To achieve this, we will carry out the following steps:

1. Define our dependency on the Rust Fibonacci number calculation package.

2. Build our calculation module with Rust.

3. Create a calculation view using Rust in our Flask application.

4. Insert Rust into our Celery task.

With this, we will have a Flask application that has a speedup due to Rust. Let's get started!

Defining our dependency on the Rust Fibonacci number calculation package

When it comes to our Rust dependency, it would be tempting to just put our Rust dependency in our `requirements.txt` file. However, this can become confusing. Also, we are using an automated process to update our `requirements.txt` file. This runs the risk of wiping our GitHub repositories from the `requirements.txt` file. We must remember that our `requirements.txt` file is just a text file. Therefore, nothing is stopping us from just adding another text file that lists our GitHub repositories and using it to install the GitHub repositories that our application is dependent on. To do this, we will populate our `src/git_repos.txt` file with the following dependency:

```
git+https://github.com/maxwellflitton/flitton-fib-rs@main
```

Now, we can install our GitHub repository dependencies with the following command:

```
pip install -r git_repos.txt
```

This will result in our Python system downloading the GitHub repository and compiling it in our Python packages. We now know which GitHub repositories are powering our application, so we can start using automation tools to update our `requirements.txt` file. Now that we have installed our Rust package, we can start building a calculation module that will use Rust.

Building our calculation model with Rust

Our calculation module will have the following structure:

```
src
├── fib_calcs
│   ├── __init__.py
│   ├── enums.py
│   └── fib_calculation.py
```

We already have our Python calculation in the `fib_calculation.py` file from the previous chapter. However, we are now supporting both Rust and Python implementations.

To do this, we will start by defining an enum in our enums.py file with the following code:

```
from enum import Enum


class CalculationMethod(Enum):


    PYTHON = "python"
    RUST = "rust"
```

With this enum, we can keep adding methods. For instance, if we were to develop microservices later on and have a separate server for calculating our Fibonacci numbers, we can add an API call to our enum and support it in our calculation interface. Depending on the configuration file, we can switch between all of them. Now that we have defined our enums, we can build our interface in the src/fib_calcs/__init__.py file:

1. First of all, we have to import what we need with the following code:

    ```
    import time
    from flitton_fib_rs.flitton_fib_rs import \
        fibonacci_number
    from fib_calcs.enums import CalculationMethod
    from fib_calcs.fib_calculation import FibCalculation
    ```

    Here, we used the time module to time how long a process takes to run. We also imported our Python and Rust implementations for our calculations. Finally, we imported our enum to map which method we used.

2. With all of this, we can start building the time process function in our src/fib_calcs/__init__.py file with the following code:

    ```
    def _time_process(processor, input_number):
        start = time.time()
        calc = processor(input_number)
        finish = time.time()
        time_taken = finish - start
        return calc, time_taken
    ```

Here, we took in a calculation function under the `processor` parameter name and passed the `input_number` parameter into the function. We also timed this process and returned it with the Fibonacci number. Now that we've done this, we can build a function that processes an input string and convert that into our enum. We will not always pass a string into our interface, but if we can load a string signaling what process type we want from a configuration file, this will be important.

3. Our processing method can be defined with the following code:

```
def _process_method(input_method):
    calc_enum = CalculationMethod._value2member_map_.
                get(input_method)
    if calc_enum is None:
        raise ValueError(
            f"{input_method} is not supported, "
            f"please choose from "
        f"{CalculationMethod._value2member_map_.keys()}")
    return calc_enum
```

Here, we can see that our string is stored in the key values of the `_value2member_map_` map. If it is not in the keys, then our enum will not support it, and the method will throw an error. However, if it exists, we return the enum associated with the key value.

4. Now, we can define the two helper functions for our interface with the following code:

```
def calc_fib_num(input_number, method):
    if isinstance(method, str):
        method = _process_method(input_method=method)

    if method == CalculationMethod.PYTHON:
        calc, time_taken = _time_process(
            processor=FibCalculation,
            input_number=input_number
        )
        return calc.fib_number, time_taken

    elif method == CalculationMethod.RUST:
        calc, time_taken = _time_process(
```

```
        processor=fibonacci_number,
        input_number=input_number
    )
    return calc, time_taken
```

Here, if we pass in a string for our method, we can convert it into an enum. If the enum points to Python, we can pass our Python calculation object, along with the input number, into our _time_process function. Then, we can return the Fibonacci number and time taken. If our enum points to Rust, we can perform the same operations but with the Rust function. With this approach, we can add and take away functionality. For instance, we can switch the timing process with another parameter that's pointing to another calculation function that does not time the process, resulting in a process that just performs the calculation without timing it if we want. However, for this example, we will be using the timing process to compare speeds. Now that we have built our interface, we can create our calculation view with this interface.

Creating a calculation view using Rust

We are hosting our view in the src/app.py file. First, we will import our interface with the following code:

```
from fib_calcs import calc_fib_num
from fib_calcs.enums import CalculationMethod
```

With this new interface and enum, we can make changes to our standard calculation view with the following code:

```
@app.route("/calculate/<int:number>")
def calculate(number):
    fib_calc = dal.session.query(FibEntry).filter_by(
        input_number=number).one_or_none()
    if fib_calc is None:
        if number < 50:
            fib_number, time_taken = calc_fib_num(
                input_number=number,
                method=CalculationMethod.PYTHON
            )
            . . .


            return f"you entered {number} " \
```

```
        f"which has a Fibonacci number of " \
        f"{fib_number} which took {time_taken}"
    . . .
```

Here, we are using the new interface. Because of this, we can also return the time taken to perform the calculation. Now, we can build our Rust calculation view. It will take the same form as the standard calculation view, meaning that you can refactor it to have the Rust and Python calculation methods in the same view based on the parameter that's passed into the URL. If not, our Rust calculation view will take the form of the following code:

```
@app.route("/rust/calculate/<int:number>")
def rust_calculate(number):
    . . .
    if fib_calc is None:
        if number < 50:
            fib_number, time_taken = calc_fib_num(
                input_number=number,
                method=CalculationMethod.RUST
            )
        . . .
```

The dots in the aforementioned code show that this is the same code that's used in the standard calculation function. Now that our Rust package has been fused with our Flask application, we can insert Rust into our Celery task.

Inserting Rust into our Celery task

When it comes to our background task in Celery, we do not have to worry about the timing. Because of the interface and configuration, we have to import the parameters and interface into the `src/task_queue/fib_calc_task.py` file with the following code:

```
from config import GlobalParams
from fib_calcs import calc_fib_num
```

With this, we can now refactor our Celery task with the following code:

```
def create_calculate_fib(input_celery):
    @input_celery.task()
    def calculate_fib(number):
```

```
        params = GlobalParams()
        fib_number, _ = calc_fib_num(input_number=number,
                                     method=params.get(
                                     "CELERY_METHOD",
                                     "rust"))
        fib_entry = FibEntry(input_number=number,
            calculated_number=fib_number)
        dal.session.add(fib_entry)
        dal.session.commit()
    return calculate_fib
```

Here, we can see that we get the global parameters. We pass the CELERY_METHOD global parameter into the params. Considering that the parameters are inherited from the dictionary class, we can use the built-in get method. We can set the default calculation method to rust if we have not defined CELERY_METHOD in the configuration file.

The application is now fully integrated, which means we can test our application. We must remember to run our development docker-compose environment, Flask application, and Celery worker. Accessing our two views will give us the following output:

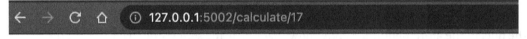

you entered 17 which has a Fibonacci number of 1597 which took 0.0008478164672851562

you entered 18 which has a Fibonacci number of 2584 which took 2.193450927734375e-05

Figure 10.1 – Flask, Python, and Rust requests

In the preceding screenshot, we can see that our Rust call is four times faster, even though the Rust request number is higher. We now have a working Python application that uses Rust to speed up the calculations. However, this is not very useful if we cannot deploy it. The internet is full of half-baked tutorials that teach you how to do something superficially in a development environment, while not being able to use or configure it in a production environment. In the next section, we will configure our Docker environment so that we can deploy our application.

Deploying Flask and Celery with Rust

For our Flask application's Docker image to support our Rust packages, we need to make some changes to the `src/Dockerfile` file. Looking at this file, we can see that our image is built on `python:3.6.13-stretch`. This is essentially a Linux environment with Python installed. When we see this, we realize that we can be confident in manipulating our Docker image environment. If we can do this in Linux, there is a high chance we can do this in our Docker image. Considering this, what we must do in our `src/Dockerfile` file is install Rust and register `cargo` with the following code:

```
. . .
RUN apt-get update -y
RUN apt-get install -y python3-dev python-dev gcc

# setup rust
RUN curl https://sh.rustup.rs -sSf | bash -s -- -y –profile
    minimal –no-modify-path

# Add .cargo/bin to PATH
ENV PATH="/root/.cargo/bin:${PATH}"
. . .
```

Luckily for us, Rust is very easy to install. Remember that the `apt-get install -y python3-dev python-dev gcc` command allows us to use compiled extensions when using Python. Now that we've done this, we can pull and compile our Rust package with the following code:

```
. . .
# Install the dependencies
RUN pip install --upgrade pip setuptools wheel
RUN pip install -r requirements.txt
RUN pip install -r git_repos.txt
. . .
```

Everything else is the same. Our image is now ready to be built with the following command while our terminal is in the root of the `src/` directory:

```
docker build . -t flask-fib
```

This will rebuild our Docker image for our Flask application. Some bits might be skipped over in this build. Don't worry – Docker caches the layers in the image build that have not been changed. This is denoted with the following printout:

```
Step 1/14 : FROM python:3.6.13-stretch
 ---> b45d914a4516
Step 2/14 : WORKDIR /app
 ---> Using cache
 ---> b0331f8a005d
Step 3/14 : ADD . /app
 ---> Using cache
```

Once a step has been changed, every step following it will be rerun since the interrupted step might change the outcome of the steps following it. Note that the build might hang when `pip` installs our Rust package. This is because the package is being compiled. You may have noticed that we have to do this every time we install the Rust package. A more optimal distribution strategy will be explored in the next chapter. Now, if we run `docker-compose` in our deployment directory, we will see that we can use our Rust Flask container without any problems.

Deploying with a private GitHub repository

If you are coding for a side project, company, or paid feature, you will be working with private GitHub repositories. This makes sense as we do not want people accessing a repository for free that you or your company plans on charging them for. However, if we set our Rust Fibonacci package's GitHub repository to private, delete all of our Flask images using the `docker image rm YOUR_IMAGE_ID_HERE` command, and run our `docker build . -t flask-fib` command again, we would get the following printout:

```
Collecting git+https://github.com/maxwellflitton/flitton-
fib-rs@main
  Running command git clone -q
https://github.com/maxwellflitton/
flitton-fib-rs /tmp/pip-req-build-ctmjnoq0
  Cloning https://github.com/maxwellflitton/flitton-fib-rs
(to revision main) to /tmp/pip-req-build-ctmjnoq0
fatal: could not read Username for 'https://github.com':
No such device or address
```

This is because our isolated Linux-based Docker image that is being built is not logged into GitHub, even though we are. As a result, the image that's being built could not pull the package from the GitHub repository. We could pass our GitHub credentials into the build via arguments, but this will show up in the image build layers. Therefore, anyone who has access to our image could look and see our GitHub credentials. This is a security hazard. Docker does have some documentation on passing in secrets. However, at the time of writing this book, the documentation is sparse and convoluted. A more straightforward approach is to clone our flitton-fib-rs package outside the image and pass it into the Docker image build, as shown here:

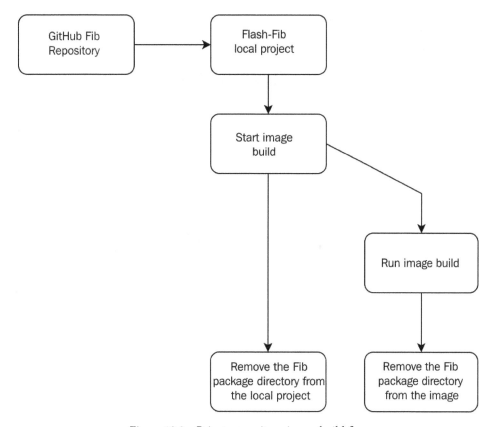

Figure 10.2 – Private repository image build flow

If we are going to use a continuous integration tool such as GitHub Actions or Travis, then we can run the process laid out in the preceding diagram with GitHub credentials passed in as secrets. GitHub Actions and Travis handle secrets with efficiency and simplicity. If we are building it locally, as we are doing in this example, then we should already be logged into GitHub as we are directly working on the Flask project in this project. To carry out the process laid out in the preceding diagram, we must carry out the following steps:

1. Build a Bash script that orchestrates the process depicted in the preceding diagram.

2. Reconfigure our Rust Fib package installment in our Dockerfile.

This is the most straightforward approach to using private GitHub repositories in our web application builds. We will start by looking at the Bash script.

Building a Bash script that orchestrates the whole process

Our script is housed in `src/build_image.sh`. First, we must declare that this is a Bash script and that the code should run in the directory of the Flask application. To do so, we must change to the directory that contains the script with the following code:

```
#!/usr/bin/env bash


SCRIPTPATH="$( cd "$(dirname "$0")" ; pwd -P )"
cd $SCRIPTPATH
```

Now, we have to clone our package and remove our `.git` file from the repository with the following code:

```
git clone https://github.com/maxwellflitton/flitton-fib-
    rs.git
rm -rf ./flitton-fib-rs/.git
```

Now, our package is just a directory. We are ready to build our Docker image. However, if we do so, it might not work because our files might be cached. To prevent this from happening, we can run our build with no cache and then remove our cloned package after the build with the following code:

```
docker build . --no-cache -t flask-fib


rm -rf ./flitton-fib-rs
```

We will have to run this script to run a build of our Flask application. However, if we were to run a build now, it would not work as our Dockerfile will still be trying to pull the directory from GitHub. To fix this, we will move on to the second step.

Reconfiguring the Rust Fib package installment in our Dockerfile

In our `src/Dockerfile` file, we must remove the `RUN pip install -r git_repos.txt` line as this will stop our image build from trying to pull from the GitHub repository. Now, we can `pip install` the local directory that has been passed in, and then remove it with the following code:

```
RUN pip install ./flitton-fib-rs
RUN rm -rf ./flitton-fib-rs
```

Now, we can build our Flask image by running the following command:

```
sh build_image.sh
```

This will result in a long printout that will eventually tell us that the image was successfully built. Running our deployment `docker-compose` file will confirm this. You may want to install our package from a different Git branch. This can be done by adding three more lines to our `src/build_image.sh` file, as shown here:

```
. . .
git clone -branch $1
   https://github.com/maxwellflitton/flitton-fib-rs.git
cd flitton-fib-rs
cd ..
rm -rf ./flitton-fib-rs/.git
. . .
```

Here, we cloned the repository containing the branch, whose name is based on the argument that's passed into the script. Once we've done this, we can remove the Git history by removing the `.git` file.

Our Rust package is now fully fused with our Python web application in Docker. One bonus of installing our Rust package when building an image is that it does not have to be compiled every time we use the image.

> **Note**
>
> We can go one step further when it comes to reducing our build, though this is optional. You do not have to do this to complete this chapter. Right now, we are installing Rust and then compiling our Rust Python package for the Fibonacci calculations. We can avoid the need to install Rust and compile every time by building wheels for a range of Linux distributions and Python versions. This can be done by pulling the ManyLinux Docker images and using them to compile our package into multiple distributions.
>
> The detailed steps on how to do this to your Python `pip` package coded in Rust are laid out in the Rust setup tools documentation (see the *Further reading* section). Once those steps are completed, you will end up with a range of wheels in your `dist` directory. Copy and paste the 3.6 version into your Flask `src` directory and instruct your Dockerfile to copy it into the image when it is being built. Once you've done this, you can point the `pip install` command directly to the wheel file you copied into the image build. The installation will be nearly instant.

While fusing Rust with our Flask application is certainly useful, since we now have a real-world example of how our Rust code can be used in a deployment setting, we can go even further. In the next section, we will have our Rust code interact with our database.

Fusing Rust with data access

In web applications, accessing a database is a big part of the process. We could import the `dal` object that we created in the `src/data_access.py` file and pass it into our Rust function, executing database operations through Python. While this will technically work, it is not ideal as we will have to waste time and effort extracting objects from the database queries, inspecting them, and converting them into Rust structs. We would then have to convert the Rust structs into Python objects before inserting them into the database. This is a lot of excess code that has a lot of interaction with Python, reducing its speed gain.

Because a database is external from the Python web application, and it contains information about its schema, we can completely bypass Python's implementations by using the `diesel` Rust crate to automatically write our schema and database models in Rust based on the live database. We can also use `diesel` to manage the connection to the database. As a result, we can directly interact with the database, reducing our reliance on Python, speeding up our code, and reducing the amount of code that we have to write. To achieve this, we have to carry out the following steps:

1. Set up our database cloning package.

2. Set up our `diesel` environment.

3. Autogenerate and configure our database models and schema.

4. Define our database connection in Rust.

5. Create a Rust function that gets all the Fibonacci records and returns them.

Once we have completed these steps, we will have a Rust package that interacts with the database that can be added to our Flask application build and used if needed. We will start by setting up our database cloning package.

Setting up our database cloning package

We should now be familiar with setting up a standard Rust package for Python. For our database package, we will have the following layout:

```
├── Cargo.toml
├── diesel.toml
├── rust_db_cloning
│   └── __init__.py
├── setup.py
├── src
│   ├── database.rs
│   ├── lib.rs
│   ├── models.rs
│   └── schema.rs
├── .env
```

You should know the role of some of these files by now. The new files have the following purposes:

- `database.rs`: Houses the function that returns a database connection
- `models.rs`: Houses the structs that define the database models, fields, and the behavior of the individual rows of a table in the database
- `schema.rs`: Houses the schema of the tables of the database
- `.env`: Houses the database URL for our **command-line interface** (**CLI**) interactions
- `Diesel.toml`: Houses the configuration for our `diesel` CLI

Now, we can turn our attention to the `setup.py` file. Looking at the package layout, you should be able to define this file yourself, and I encourage you to give it a try. Here is an example of the barebones `setup.py` file that is needed to enable this package to be installed with `pip`:

```python
#!/usr/bin/env python
from setuptools import dist
dist.Distribution().fetch_build_eggs(['setuptools_rust'])
from setuptools import setup
from setuptools_rust import Binding, RustExtension

setup(
    name="rust-database-cloning",
    version="0.1",
    rust_extensions=[RustExtension(
        ".rust_db_cloning.rust_db_cloning",
        path="Cargo.toml", binding=Binding.PyO3)],
    packages=["rust_db_cloning"],
    zip_safe=False,
)
```

With this, our `rust_db_cloning/__init__.py` file contains the following code:

```python
from .rust_db_cloning import *
```

Now, we can move onto our `Cargo.toml` file, which will list some dependencies that you are familiar with, as well as the new `diesel` dependency:

```
[package]
name = "rust_db_cloning"
version = "0.1.0"
authors = ["maxwellflitton"]
edition = "2018"


[dependencies]
diesel = { version = "1.4.4", features = ["postgres"] }
dotenv = "0.15.0"


[lib]
name = "rust_db_cloning"
crate-type = ["cdylib"]


[dependencies.pyo3]
version = "0.13.2"
features = ["extension-module"]
```

With that, we have defined the basics for our package to be installed via `pip`. It will not be installed yet as we have nothing in our `src/lib.rs` file, but we will fill that file out in the final step. Now, we can move on to the next step, which is setting up our `diesel` environment.

Setting up the diesel environment

We will be cloning our schema from our development database so that we can hardcode the URL into our `.env` file, as follows:

```
DATABASE_URL=postgresql://user:password@localhost:5432/fib
```

Since this database configuration will never end up in a production environment and is merely used to generate the schema and models from a development database, it is OK if this URL gets into the wrong hands. Having this hardcoded into your GitHub repository is not the end of the world. With this in mind, we can define where we want our schema to be printed in our `diesel.toml` file with the following code:

```
[print_schema]
file = "src/schema.rs"
```

Now that we have written everything we need, we can start installing and running the `diesel` CLI. You may get compilation errors when installing and compiling `diesel`. If this is the case while you are reading this, you can get around these compilation errors by switching to Rust `nightly`. Rust `nightly` provides the latest releases of Rust; however, it is less stable. Therefore, you should try and follow these steps without switching to `nightly` but if you find that you need to, then you can switch to `nightly` by installing it with the following code:

```
rustup toolchain install nightly
```

Once it has been installed, we can switch to `nightly` with the following command:

```
rustup default nightly
```

Your Rust compilations will be running in `nightly` now. Going back to setting up our `diesel` environment, we have to install the `diesel` CLI with the following command:

```
cargo install diesel_cli --no-default-features
--features postgres
```

With this, we can now use the CLI combined with the URL in the `.env` file to interact with our database.

Autogenerating and configuring our database models and schema

In this step, we will be interacting with the development database in Docker. Considering this, before moving on, you need to open another terminal and run the development `docker-compose` environment in the `flask-fib` repository. Running this will run the database that we will connect to so that we can access the database schema and models. Now that the CLI has been installed, we can print the schema with the following command:

```
diesel print-schema > src/schema.rs
```

There will be no printouts in the terminal but if we open our `src/schema.rs` file, we will see the following code:

```
table! {
    alembic_version (version_num) {
        version_num -> Varchar,
    }
}
table! {
    fib_entries (id) {
        id -> Int4,
        input_number -> Nullable<Int4>,
        calculated_number -> Nullable<Int4>,
    }
}
allow_tables_to_appear_in_same_query!(
    alembic_version,
    fib_entries,
);
```

Here, we can see that our `alembic` version is in the schema as a separate table. This is how `alembic` keeps track of the migrations. We can also see that our `fib_entries` table has been mapped. While we could have done this ourselves without the `diesel` CLI, it is a lifesaver, ensuring that the schema is always up to date with the database. This also saves time in big, complex databases and reduces errors.

Now that our schema has been defined, we can define our models with the following command:

```
diesel_ext > src/models.rs
```

This gives us the following code:

```
#![allow(unused)]
#![allow(clippy::all)]


#[derive(Queryable, Debug, Identifiable)]
#[primary_key(version_num)]
pub struct AlembicVersion {
    pub version_num: String,
}
#[derive(Queryable, Debug)]
pub struct FibEntry {
    pub id: i32,
    pub input_number: Option<i32>,
    pub calculated_number: Option<i32>,
}
```

This is not completely perfect, and we have to make some changes. The models do not have the tables defined. diesel assumes that the table name is just the plural of the model's name. For instance, if we have a data model called *test*, then diesel would assume that the table is called *tests*. However, for us, this is not the case as we specifically defined our tables in our Flask application when running migrations in the previous chapter. We can also remove the two allow macros as we will not be using this functionality. Instead, we will import our schemas and define them in the table macro. After this rearrangement, our src/models.rs file should look like this:

```
use crate::schema::fib_entries;
use crate::schema::alembic_version;


#[derive(Queryable, Debug, Identifiable)]
#[primary_key(version_num)]
#[table_name="alembic_version"]
pub struct AlembicVersion {
    pub version_num: String,
```

```
}
#[derive(Queryable, Debug, Identifiable)]
#[table_name="fib_entries"]
pub struct FibEntry {
    pub id: i32,
    pub input_number: Option<i32>,
    pub calculated_number: Option<i32>,
}
```

Our models and schema are now ready to be used in our Rust package. Considering this, we can move on to the next step, which is defining our database connection.

Defining our database connection in Rust

Our database connection would traditionally take the database URL from the environment and use this to make a connection. However, this is a Rust package that is an appendage to our Flask application. There is no point in having another sensitive piece of information that has to be loaded. Therefore, to avoid extra complications and another point of security failure, we will merely pass the database URL from the Flask application to make the connection, since the Flask application is managing the configuration and loading the sensitive data anyway. The entirety of our database connection can be handled in our `src/database.rs` file. First, we must import what we need with the following code:

```
use diesel::prelude::*;
use diesel::pg::PgConnection;
```

`prelude` helps us use the `diesel` macros, and `PgConnection` is what we will return to get a database connection. With this, we can build our database connection function with the following code:

```
pub fn establish_connection(url: String) -> PgConnection {
    PgConnection::establish(&url)
        .expect(&format!("Error connecting to {}", url))
}
```

This can be imported anywhere where we need a database connection. At this point, we can start creating a function that gets all the records and returns them in dictionaries.

Creating a Rust function that gets all the Fibonacci records and returns them

To avoid excessive complexity in this example, we will be doing everything in the `src/lib.rs` file. However, it is advised that you build some modules and import them into the `src/lib.rs` file for more complex packages. First of all, we will import everything we need to build the function and bind it with the following code:

```
#[macro_use] extern crate diesel;
extern crate dotenv;
use diesel::prelude::*;

use pyo3::prelude::*;
use pyo3::wrap_pyfunction;
use pyo3::types::PyDict;

mod database;
mod schema;
mod models;

use database::establish_connection;
use models::FibEntry;
use schema::fib_entries;
```

The order of the imports matters here. We import the `diesel` crate with macro use straightaway. Therefore, files such as `database` and `schema` will not error out because they are using `diesel` macros. `dotenv` is not being used in our example as we are passing in the database URL from the Python system. However, it's useful to know about this if you want to get database URLs from the environment. Then, we can import the `pyo3` macros and structs that we need, and the structs and functions that we defined. With these imports, we can define our `get_fib_entries` function with the following code:

```
#[pyfunction]
fn get_fib_entries(url: String, py: Python) -> Vec<&PyDict>
{

    let connection = establish_connection(url);
```

```
    let fibs = fib_entries::table
        .order(fib_entries::columns::input_number.asc())
        .load::<FibEntry>(&connection)
        .unwrap();

    let mut buffer = Vec::new();

    for i in fibs {
        let placeholder = PyDict::new(py);
        placeholder.set_item("input number",
            i.input_number.unwrap());
        placeholder.set_item("fib number",
            i.calculated_number.unwrap());
        buffer.push(placeholder);
    }
}
```

Using Python to build a list of dictionaries is not new, and neither is the definition of the function. What is new, however, is establishing a connection, ordering it using the schema columns, and loading it as a list of `FibEntry` structs. We pass a reference to the connection into our query and unwrap it as it returns a result. We can chain more functions to it, such as `.filter`, if needed. The `diesel` documentation does a good job of covering the different types of queries and inserts you can perform. Once we've done this, we can add it to our `rust_db_cloning` module with the following code:

```
#[pymodule]
fn rust_db_cloning(py: Python, m: &PyModule)
    -> PyResult<()> {
    m.add_wrapped(wrap_pyfunction!(get_fib_enteries));
    Ok(())
}
```

With this, our code is ready to be uploaded to the GitHub repository and used in our Flask application.

Now, we can quickly test whether our package works before defining it in our Dockerfile. First of all, we need to `pip install` it in our Flask application virtual environment. This is another point where you might have compilation issues. To get around this, you might have to switch to Rust `nightly` to `pip install` the package you just built. Once our package has been installed, we can check it out by adding a simple `get` view to our Flask application. In the `src/app.py` file of our Flask application, we can import our function with the following code:

```
from rust_db_cloning import get_fib_entries
```

Now, we can define our `get` view with the following code:

```
@app.route("/get")
def get():
    return str(get_fib_entries(dal.url))
```

Remember that in the previous chapter, we defined the `url` attribute of `dal` with the URL from `GlobalParams`, which was loaded from the `.yml` config file. We have to turn it into a string; otherwise, the Flask serialization will not be able to process it. Running this in the development `docker-compose` environment will give us the following output:

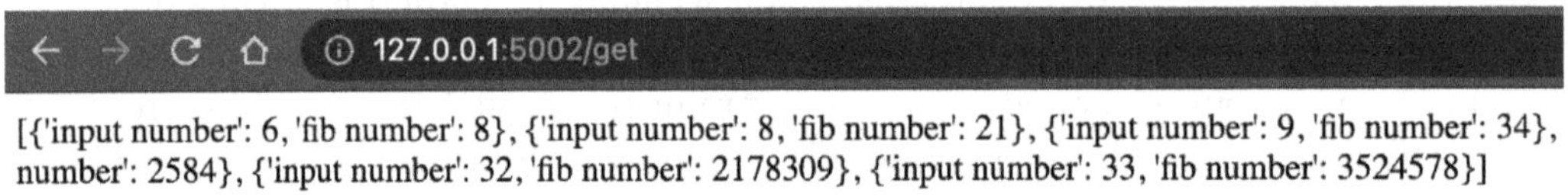

[{'input number': 6, 'fib number': 8}, {'input number': 8, 'fib number': 21}, {'input number': 9, 'fib number': 34}, number': 2584}, {'input number': 32, 'fib number': 2178309}, {'input number': 33, 'fib number': 3524578}]

Figure 10.3 – Simple get view from our Flask application

You may have different numbers, depending on what you have in your database. However, what we have here is a Rust package that keeps up with the changes in the database that can interact directly with the database. Now that this is working in our development setup, we can start packaging our Rust `nightly` package for deployment.

Deploying Rust nightly in Flask

To package our `nightly` database Rust package so that it can be deployed, we have to add another clone of our GitHub repository to our build Bash script, install `nightly`, and switch to it when we are installing our database package with `pip`. You can probably guess what we are going to achieve by cloning our database GitHub repository in our Bash script.

For reference, our `src/build_image.sh` file will take the form of the following code:

```
. . .
git clone https://github.com/maxwellflitton/
flitton-fib-rs.git
git clone https://github.com/maxwellflitton/
rust-db-cloning.git

rm -rf ./flitton-fib-rs/.git
rm -rf ./rust-db-cloning/.git

docker build . --no-cache -t flask-fib

rm -rf ./flitton-fib-rs
rm -rf ./rust-db-cloning
```

Here, we can see that we have merely added the code for cloning the `rust-db-cloning` repository, removed the `.git` file in that `rust-db-cloning` repository, and then removed the `rust-db-cloning` repository once the image build has finished. When it comes to our Dockerfile, these steps will remain the same. The only difference is that after installing our normal Rust package, we install `nightly`, switch to it, and then install our database package. This can be achieved with the following code:

```
. . .
RUN pip install ./flitton-fib-rs
RUN rm -rf ./flitton-fib-rs

RUN rustup toolchain install nightly
RUN rustup default nightly

RUN pip install ./rust-db-cloning
RUN rm -rf ./rust-db-cloning
. . .
```

Even though one is compiled with normal Rust, while the other is compiled with Rust `nightly`, they will both run fine when the application is running. Building this image and running it in the deployment `docker-compose` environment will show us that the container will process the Rust computation view and get it from the database view without any problems. With this, we now have all the tools we need to fuse Rust into Python web applications and deploy them in Docker.

Summary

In this chapter, we have put all our Rust fusing skills to work to build packages that are baked into Docker images for a Python web application. We attached Rust packages directly to the web application, and then to the Celery worker, resulting in a significant speedup when we asked our web application to calculate the Fibonacci number.

Then, we altered our build process to take Rust packages from private GitHub repositories when building our Python web application image. Finally, we connected directly to the database with Rust and used Rust `nightly` to compile it. We managed to include this in our Python web application Docker build. This resulted in us not only being able to fuse Rust into a deployable web application but also use Rust `nightly` and databases to solve our problems

With this in mind, we can now use what we have learned in this book for production web applications. You can now start coding in Rust and plug your Rust packages into existing Python web applications that can be deployed in Docker, without having to make major changes to the Python web application build process.

Reaching for Rust to solve a speed bottleneck or to ensure that the code is consistent and safe in a live Python web application is something you can do in your day job tomorrow. You can now bring forward the fastest memory-safe programming language into your Python projects without having to overhaul your existing system. You are now capable of bridging the gap between practically maintaining an existing tried and tested system and a cutting-edge language. In the next and final chapter, we will cover some best practices. But right now, you know the key concepts to go and change your project or organization.

Questions

1. How does directly connecting to a database in Rust reduce code?
2. Why can't we just pass login credentials into our Docker image build Dockerfile?
3. We did not make any migrations in this chapter. How did we map our models and schema of a database to our Rust module, and how do we continue to keep up with database changes?

4. Why do we pass the database URL into our database Rust package as opposed to loading it from a config file or environment?

5. Do we have to do anything extra if we were going to fuse Rust with a Django, bottle, or FastAPI Python web application?

Answers

1. Directly connecting to a database with Rust reduces the amount of code we must write as we do not have to inspect the Python objects that are returned from the Python database call. We also do not have to package data into Python objects before inserting them into the database. This essentially removes a whole layer of code that we must write when interacting with the database.

2. If someone gets hold of our image, they can access the layers of the build. As a result, they can access the arguments that have been passed into the build. This will mean that they can see the credentials we use to log in.

3. We used the `diesel` crate to connect to the database and automatically print the schemas and models based on the database it connects to. We can do this repeatedly to keep up to date with new database migrations.

4. We must remember that our Rust database package is an appendage to our Python web application. Our Python web application has already loaded the database URL. Loading credentials into our package just adds another possibility for a security breach with no advantages.

5. No – our fusing method is completely detached from the `pip` installation process and the database mapping process.

Further reading

- Diesel documentation for Rust (2021): Crate Diesel: `https://diesel.rs`

- Setup tools Rust documentation (2021): Distributing a Rust Python package with wheels: `https://pypi.org/project/setuptools-rust/`

- ManyLinux GitHub (2021): `https://github.com/pypa/manylinux`

11

Best Practices for Integrating Rust

In *Chapter 10, Injecting Rust into a Python Flask App*, we managed to fuse our Rust code with a Python web application. In this final chapter, we will wrap up what we have covered in the book with best practices. These practices are not essential to fuse Rust with Python; however, they will help us in avoiding pitfalls when building bigger packages in Rust. When it comes to best practices, we can Google search the topic *SOLID principles*, which will give us loads of free information on how to keep code generally clean. But instead of regurgitating these principles, we will cover concepts that are specific to using Rust and Python together. We will learn how to keep the Rust/Python implementation as simple as possible if the requirements are not too demanding. We will also understand what Python and Rust excel in when it comes to computational tasks and Python interfaces. We then investigate traits in Rust and how they can help us organize and structure our structs. Finally, we look into keeping it simple when we want data parallelism with the Rayon crate.

In this chapter, we will cover the following topics:

- Keeping our Rust implementation simple by piping data to and from Rust
- Giving the interface a native feel with objects
- Using traits as opposed to objects
- Keeping data-parallelism simple with Rayon

Covering these topics will enable us to avoid pitfalls when building bigger packages that are more complex. We will also be able to build Rust solutions for smaller projects faster as we will learn that we do not have to rely on Python setup tools and installation with `pip`.

Technical requirements

The code and data for this chapter can be found at the following link:

`https://github.com/PacktPublishing/Speed-up-your-Python-with-Rust/tree/main/chapter_eleven`

Keeping our Rust implementation simple by piping data to and from Rust

We have covered everything we need to integrate Rust into our Python system. We can build Rust packages that can be installed using `pip` and use them in Docker when integrating with a web application. However, reaching for a setup tool can be too much effort if the problem being solved is small and simple. For instance, if in a situation we were merely opening a **comma-separated values (CSV)** file full of numbers in Python, calculating the Fibonacci numbers, and then writing them in another file, then it would make sense to just write the program in Rust. However, we do not start building a Rust package with Python setup tools if we have a more complicated Python standalone script that just needs a simple speedup with Rust—it is still just a standalone script. Instead, we pipe data. This means we pass data from a Python script to a Rust standalone binary and back to a Python script for computing Fibonacci numbers, as shown here:

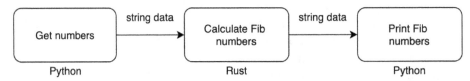

Figure 11.1 – Process of a basic pipeline

To achieve the same speed as the Rust Fibonacci calculation package without having to use any setup tools, we must carry out the following steps:

1. Build a Python script that formulates the numbers for calculation.

2. Build a Rust file that accepts the numbers, calculates the Fibonacci numbers, and returns the calculated numbers.

3. Build another Python script that accepts the calculated numbers and prints them out.

With this, we will be able to have a simple pipeline. While each file is isolated and we can build in any order, it makes sense to start with *Step 1*.

Building a Python script that formulates the numbers for calculation

For this example, we will just hardcode the input numbers that we are passing into our pipeline, but nothing is stopping you from reading your data from files or taking in numbers from command-line arguments. In our `input.py` file, we can write to `stdout` with the following code:

```python
import sys

sys.stdout.write("5\n")
sys.stdout.write("6\n")
sys.stdout.write("7\n")
sys.stdout.write("8\n")
sys.stdout.write("9\n")
sys.stdout.write("10\n")
```

With this, if we run this script with the Python interpreter, we get the following output:

```
$ python input.py
5
6
7
8
9
10
```

With this, we can now move on to the next step.

Building a Rust file that accepts the numbers, calculates the Fibonacci numbers, and returns the calculated numbers

For our Rust file, we must have everything contained in the file to keep it as simple as possible. We can span it over multiple files if needed, but for a simple calculation, keeping it all in one file is good enough. In our `fib.rs` file, we initially import what we need and define our Fibonacci function with the following code:

```
use std::io;
use std::io::prelude::*;


pub fn fibonacci_reccursive(n: i32) -> u64 {
    match n {
        1 | 2 => 1,
        _     => fibonacci_reccursive(n - 1) +
                fibonacci_reccursive(n - 2)
    }
}
```

Here, we can see that there is nothing new; we are merely going to use `std::io` to get the numbers piped into the file, and then calculate the Fibonacci number, sending it to the next file in the pipeline with the following `main` function:

```
fn main() {
    let stdin = io::stdin();
    let stdout = std::io::stdout();
    let mut writer = stdout.lock();

    for line in stdin.lock().lines() {
        let input_int: i32 = line.unwrap().parse::<i32>() \
            .unwrap();
        let fib_number = fibonacci_reccursive(input_int);
        writeln!(writer, "{}", fib_number);
    }
}
```

Here, we can see that we define `stdin` to receive the numbers sent to the Rust program, and `stdout` to send out the calculated Fibonacci numbers. We then loop through the lines sent into the Rust program and then parse each line at an integer. We then calculate the Fibonacci number and then send it using the macro that we imported with the `io` prelude. With this, we can now compile our Rust file with the following command:

```
rustc fib.rs
```

This will compile our Rust file. We can now run both files, piping the data from the Python file to the compiled Rust code with the following command:

```
$ python input.py | ./fib
5
8
13
21
34
55
```

Here, we can see that the numbers from the `python input.py` command get piped into the Rust code returning the calculated Fibonacci numbers. With this, we can now move on to the final step, where we get the calculated Fibonacci numbers from the Rust code and print them out.

Building another Python script that accepts the calculated numbers and prints them out

Our `output.py` file is very straightforward. It takes the following form:

```python
import sys


for i in sys.stdin.readlines():
    try:
        processed_number = int(i)
        print(f"receiving: {processed_number}")
    except ValueError:
        pass
```

We have a `try` block because the start and the end of the data passed into the last Python script will have empty lines, and they will fail when we try to convert them into integers. We then print out the data with `"receiving: {processed_number}"` added to the last script to make it clear that it is the `output.py` file printing out the numbers. This gives us the printout, as follows:

```
$ python input.py | ./fib | python output.py
receiving: 5
receiving: 8
receiving: 13
receiving: 21
receiving: 34
receiving: 55
```

We can time how long it takes for the pipeline to run using the `time` command. If we compare this to pure Python with the example numbers that we have used, pure Python will be faster. However, we know that Rust is much faster than pure Python code. Because the **input/output** (**I/O**) operations take time, it is not worth implementing the pipeline if you are going to calculate one or two small numbers. However, to demonstrate the value of our approach, we can write the following pure Python code in the `pure_python.py` file:

```python
def recur_fib(n: int) -> int:
    if n <= 2:
        return 1
    else:
        return (recur_fib(n - 1) +
                recur_fib(n - 2))


for i in [5, 6, 7, 8, 9, 10, 15, 20, 25, 30]:
    print(recur_fib(i))
```

This gives us the following printout:

```
$ time python pure_python.py
5
8
13
21
```

```
34
55
610
6765
75025
832040

real    0m0.315s
user    0m0.240s
sys    0m0.027s
```

Adding the same numbers to the pipeline gives us the following printout:

```
$ time python input.py | ./fib | python output.py
receiving: 5
receiving: 8
receiving: 13
receiving: 21
receiving: 34
receiving: 55
receiving: 610
receiving: 6765
receiving: 75025
receiving: 832040

real    0m0.054s
user    0m0.050s
sys    0m0.025s
```

Here, we can see that our pipeline is much faster. The gap between Rust and pure Python will just get larger as the numbers increase and become more numerous. We can see here that this is a lot easier with fewer moving parts. If our program is simple, then we keep the construction and use of Rust simple.

Giving the interface a native feel with objects

Python is an object-oriented language. When we are building our Rust packages, we need to keep the friction of adoption low. The adoption of Rust packages would be better if we keep our interfaces as objects. Most Python packages have object interfaces. Calculations are done with inputs, and the Python object has a range of functions and attributes that give us the results of those calculations. While we did cover creating classes in Rust with pyo3 macros in *Chapter 6, Working with Python Objects in Rust*, in the *Constructing our custom Python objects in Rust* section, it is advised that we understand the pros and cons of doing this. We remember that classes written in Rust are faster. However, the freedom of inheritance and metaclassing with pure Python is useful. As a result, it is best to leave the construction and organization of the object interface in pure Python. Any calculations that need to be done can be done in Rust. To demonstrate this, we can use the simple physics example of a particle's **two-dimensional (2D)** trajectory, as seen in the following screenshot:

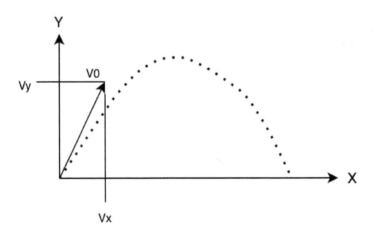

Figure 11.2 – Simple 2D physics trajectory

Here, we can see that the initial velocity is denoted as *V0*. The projection on the *x* axis is denoted by *Vx*, and the projection on the *y* axis is denoted by *Vy*. The dashed line is the particle at every point in time. Time here is another dimension. Our equations for each position in time and the endpoint in time (when it hits the ground) are defined as follows:

$$y(t) = V_y t - \frac{1}{2} g\, t^2$$

$$x(t) = V_x t$$

$$t_{end} = \frac{2V_y}{g}$$

Here, g is the constant for gravity. We also want to know the position of the particle at a certain point in time. We do this by calculating the last point in time, then looping through all the time points between zero and the last point in time, calculating the x and y coordinates. All we would need is the initial velocity in x and y. The loop through with the position calculations would all be done in Rust. The dictionary where all the keys are all the times and all the values are tuples of x and y is housed in the Python object. We could write one function that processes all the times and returns a dictionary in Rust called `calculate_coordinates`. Using it in our Python class would look like this:

```python
from rust_package import calculate_coordinates


class Particle:
    def __init__(self, v_x, v_y):
        self.co_dict = calculate_coordinates(v_x, v_y)

    def get_position(self, time) -> tuple:
        return self.co_dict[time]
    @property
    def times(self):
        return list(self.co_dict.keys())
```

The user would just have to import the `Particle` object, initialize it with the initial velocity in the x and y coordinates, and then input times to get the coordinates. To plot all the positions for a particle, we would use our class with the following code:

```python
from . . . import Particle


particle = Particle(20, 30)


x_positions = []
y_positions = []


for t in particle.times:
    x, y = particle.get_position(t)
```

```
        x_positions.append(x)
        y_positions.append(y)
```

This is intuitive to Python. We have kept all our number crunching in Rust to get that speed, but we have managed to keep all of our interfaces, including accessing times and positions in Python. As a result, a developer using this package would not know that it is written in Rust—they would just appreciate that it is fast. We can drive home the benefits of keeping the formation and access to data 100% in Python while doing all the calculations in Rust with metaclassing.

In our particle system, we could load a lot of data for the initial velocities of particles. As a result, our system would calculate the trajectories for a range of particles that we load. However, if we load two particles with the same initial velocities, they will both have the same trajectories. Considering this, it would not make sense for us to calculate the trajectory for both particles. To avoid this, we do not need to store anything in a file or database for reference; we just need to implement a flyweight design pattern. This is where we check the parameters passed into the object. If they are the same as the previous instance, we just return the previous instance. The flyweight pattern is defined with the following code:

```
class FlyWeight(type):

    _instances = {}

    def __call__(cls, *args, **kwargs):
        key = str(args[0]) + str(args[1])
        if key not in cls._instances:
            cls._instances[key] = super( \
                FlyWeight, cls).__call__(*args, **kwargs)
        return cls._instances[key]
```

Here, we can see that we combine the initial velocities to define a key, and then check to see if there is already an instance with these velocities. If there is, we return the instance from our _instances dictionary. If not, we create a new instance and insert it into our _instances dictionary. Our particles will then take the form of the following code:

```
class Particle(metaclass=FlyWeight):

    def __init__(self, v_x, v_y):
        self.co_dict = calculate_coordinates(v_x, v_y)
```

```python
    def get_position(self, time) -> tuple:
        return self.co_dict[time]


    @property
    def times(self):
        return list(self.co_dict.keys())
```

Here, our particles will now adhere to the flyweight pattern. We can test this with the following code:

```python
test = Particle(4, 6)
test_two = Particle(3, 8)
test_three = Particle(4, 6)


print(id(test))
print(id(test_three))
print(id(test_two))
```

Running this will give us the following printout:

```
140579826787152
140579826787152
140579826787280
```

Here, we can see that the two particles that have the same initial velocities have the same memory address, so it works.

We can initialize these particles anywhere and this design pattern will apply, ensuring that we do not perform duplicate calculations. Considering that we are writing Python extensions in Rust, the flyweight pattern really shows us how much control we get with how the interface is called, used, and displayed. Even though we have built our interfaces in Python, this does not mean that we do not have to structure our Rust code. This brings us on to the next section, in which we discuss how to lean into traits as opposed to objects when it comes to structuring our Rust code.

Using traits as opposed to objects

As a Python developer, it is tempting to build structs that inherit via the composition of other structs. **Object-oriented programming (OOP)** is well supported in Python; however, there are many reasons why Rust is favored, and one of them is traits. As seen in the following screenshot, traits enable us to separate data from behavior:

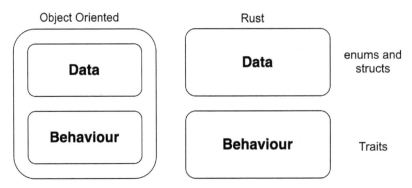

Figure 11.3 – Difference between traits and objects

This gives us a lot of flexibility as the data and behavior are decoupled, enabling us to slot behavior in and out of structs as we need. Structs can have a portfolio of traits without giving us disadvantages arising from multiple inheritance. To demonstrate this, we are going to create a basic doctor, patient, nurse program so that we can see how different structs can have different traits, allowing them to move through functions. We are going to see how traits affect the way we lay out code over multiple files. Our program will have the following layout:

With this structure, the flow of our code will take the following form:

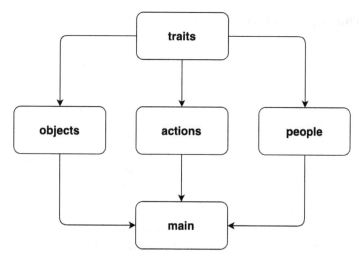

Figure 11.4 – Code flow of a simple trait-based program

With this, we can see that our code is decoupled. Our traits go into all the other files to define the behavior of those files. To build this program, we must carry out the following steps:

- Defining traits—building traits for our structs
- Defining struct behavior with traits
- Passing traits through functions
- Storing structs with common traits
- Running our program in the main.rs file.

With this, we can start by defining our traits in the first subsection.

Defining traits

Before we start defining traits, we must conceptualize the types of people in our program that we are defining behavior for. They are laid out as follows:

- **Patient**: This person does not have any clinical skills, but actions are performed on them.
- **Nurse**: This person has clinical skills but cannot prescribe or diagnose.
- **Nurse practitioner**: This person has clinical skills and can prescribe but cannot diagnose.

- **Advanced nurse practitioner**: This person has clinical skills and can prescribe and diagnose.

- **Doctor**: This person has clinical skills and can prescribe and diagnose.

What we can see here is that they are all humans. Therefore, they are all able to speak and introduce themselves. So, in our `traits.rs` file, we can create a `Speak` trait with the following code:

```
pub trait Speak {
    fn introduce(&self) -> ();
}
```

If a struct implements this trait, it will have to create its own `introduce` function with the same return and input parameters. We can also see that everyone apart from the patient has clinical skills. To accommodate this, we can implement a clinical skills trait with the following code:

```
pub trait ClinicalSkills {
    fn can_prescribe(&self) -> bool {
        return false
    }
    fn can_diagnose(&self) -> bool {
        return false
    }
    fn can_administer_medication(&self) -> bool {
        return true
    }
}
```

Here, we can see that we have defined the most common attributes for each clinician. Only two people—the doctor and the advanced nurse practitioner—can diagnose and prescribe. However, all of them can administer medication. We can implement this trait for all clinicians and then overwrite specifics. We must note that because the doctor and advanced nurse practitioner have the same possibilities in terms of diagnosing and prescribing, we can create another trait for this to prevent repeating ourselves, and a trait for the patient with the following code:

```
pub trait AdvancedMedical {}
pub trait PatientRole {
```

```
    fn get_name(&self) -> String;
}
```

We have now defined all the traits that we need. We can start using them to define our people in the next subsection.

Defining struct behavior with traits

Before we define any structs, we must import our traits into our `people.rs` file with the following code:

```
use super::traits;
use traits::{Speak, ClinicalSkills, AdvancedMedical, \
    PatientRole};
```

We now have all our traits, so we can define our people in the program with the following code:

```
pub struct Patient {
    pub name: String
}
pub struct Nurse {
    pub name: String
}
pub struct NursePractitioner {
    pub name: String
}
pub struct AdvancedNursePractitioner {
    pub name: String
}
pub struct Doctor {
    pub name: String
}
```

Sadly, there is some repetition here. This is also going to happen with our Speak trait; however, it is important to keep these structs separate as we will slot traits into them later, so we need them to be decoupled. We can implement our Speak trait for each person with the following code:

```
impl Speak for Patient {
    fn introduce(&self) {
        println!("hello I'm a Patient and my name is {}", \
            self.name);
    }
}
impl Speak for Nurse {
    fn introduce(&self) {
        println!("hello I'm a Nurse and my name is {}", \
            self.name);
    }
}
impl Speak for NursePractitioner {
    fn introduce(&self) {
        println!("hello I'm a Practitioner and my name is \
            {}", self.name);
    }
}
. . .
```

We can continue this pattern and implement Speak traits for all the people structs. Now this is done, we can implement our clinical skills and patient role traits for our people with the following code:

```
impl PatientRole for Patient {
    fn get_name(&self) -> String {
        return self.name.clone()
    }
}
impl ClinicalSkills for Nurse {}

impl ClinicalSkills for NursePractitioner {
    fn can_prescribe(&self) -> bool {
```

```
        return true
    }
}
```

Here, we can see that our people structs have the following traits:

- The `Patient` struct has the standard `PatientRole` trait.

- The `Nurse` struct has the standard `ClinicalSkills` trait.

- The `NursePractitioner` struct has the standard `ClinicalSkills` trait with the `can_prescribe` function overwritten to return `true`.

Now that we have our clinical skills applied to our standard clinicians, we can now apply our advanced traits with the following code:

```
impl AdvancedMedical for AdvancedNursePractitioner {}
impl AdvancedMedical for Doctor {}

impl<T> ClinicalSkills for T where T: AdvancedMedical {
    fn can_prescribe(&self) -> bool {
        return true
    }
    fn can_diagnose(&self) -> bool {
        return true
    }
}
```

Here, we apply the `AdvancedMedical` trait to our `Doctor` and `AdvancedNursePractitioner` structs. However, we know that these structs are also clinicians. We need them to have clinical skills. Therefore, we implement `ClinicalSkills` for the `AdvancedMedical` trait. We then overwrite the `can_prescribe` and `can_diagnose` functions to `true`. Therefore, doctors and advanced nurse practitioners have both `ClinicalSkills` and `AdvancedMedical` traits and can diagnose and prescribe. With this, our people structs are ready to be passed into functions. We will do this in the next subsection.

Passing traits through functions

To perform actions such as updating a database or sending data to a server, we are going to pass our people structs through functions where clinicians can act on patients. To do this, we must import our traits in our `actions.rs` file with the following code:

```
use super::traits;
use traits::{ClinicalSkills, AdvancedMedical, PatientRole};
```

Our first action is to admit a patient. This can be done by anyone with clinical skills. Considering this, we can define this action with the following code:

```
pub fn admit_patient<Y: ClinicalSkills>(
    patient: &Box<dyn PatientRole>, _clinician: &Y) {
    println!("{} is being admitted", patient.get_name());
}
```

Here, we can see that our clinician being passed in is anything with a `ClinicalSkills` trait, which means all our clinician structs. However, it must be noted that we are also passing in `&Box<dyn PatientRole>` for the patient. This is because we will be using a patient list when passing in patients. We can have multiple patients assigned to a clinician. We will explore why we are using `&Box<dyn PatientRole>` in the next subsection when we define our patient list struct. The next action is to diagnose a patient, which is defined with the following code:

```
pub fn diagnose_patient<Y: AdvancedMedical>(
    patient: &Box<dyn PatientRole>, _clinician: &Y) {
    println!("{} is being diagnosed", patient.get_name());
}
```

Here, it makes sense to have the `AdvancedMedical` trait to diagnose. If we try to pass in a `Nurse` or `NursePractitioner` struct, the program will fail to compile due to mismatching traits. We can then have a prescribe medication action, which takes the form of the following code:

```
pub fn prescribe_meds<Y: ClinicalSkills>(
    patient: &Box<dyn PatientRole>, clinician: &Y) {
    if clinician.can_prescribe() {
        println!("{} is being prescribed medication", \
            patient.get_name());
    } else {
```

```
        panic!("clinician cannot prescribe medication");
    }
}
```

Here, we can see that the `ClinicalSkills` trait is accepted but the code will throw an error if the clinician cannot prescribe. This is because our `NursePractitioner` struct can prescribe. We could also make a third intermediate trait and apply it to doctor, advanced, and normal nurse practitioners. However, this is just one check as opposed to implementing a new trait for all three clinician structs. Our last action is that of administering medication and discharging the patient, which can be done by all our clinician structs; therefore, it takes the following form:

```
pub fn administer_meds<Y: ClinicalSkills>(
    patient: &Box<dyn PatientRole>, _clinician: &Y) {
    println!("{} is having meds administered", \
      patient.get_name());
}
pub fn discharge_patient<Y: ClinicalSkills>(
    patient: &Box<dyn PatientRole>, _clinician: &Y) {
    println!("{} is being discharged", patient.get_name());
}
```

We can now pass our people structs through a range of actions, with our compiler refusing to compile if we pass through the wrong person struct into the function. In the next subsection, we will be storing structs with traits in a patient list.

Storing structs with common traits

When it comes to a patient list, it is tempting to just store patient structs in a vector. However, this does not give us much flexibility. For instance, let's say that our system is deployed, and one of the nurses in the hospital is sick and must be admitted. We could allow this by implementing the `PatientRole` trait to the `Nurse` struct without having to rewrite anything else. We might also need to expand the different types of patients, adding more structs such as `ShortStayPatient`, or `CriticallySickPatient`. Because of this, we store our patients with the `PatientRole` trait in our `objects.rs` file with the following code:

```
use super::traits;
use traits::PatientRole;
```

```
pub struct PatientList {
    pub patients: Vec<Box<dyn PatientRole>>
}
```

We must wrap our structs in a box because we do not know the size at compile time. Different structs of different sizes can implement the same trait. A Box is a pointer on the heap memory. Because we know the size of pointers, we know the size of memory being added to the vector at compile time. The dyn keyword is used to define that it is a trait that we are referring to. Managing to access the struct directly in the patients vector is not going to happen, as again, we do not know the size of the struct. Therefore, we access the data of the struct via the get_name function in the PatientRole trait in our action functions. Traits are also pointers. We can still build functions such as constructors for our struct. However, when it comes to our Patient struct being passed through an action function that we created, our PatientRole trait acts as an interface between our Patient struct and our admit_function function. We now have everything we need, so we can move on to our next subsection to put it all together and run it in our main.rs file.

Running our traits in the main file

Running all our code together is straightforward and safe. Here's what we need to do:

1. First, we import all we need in our main.rs file with the following code:

```
mod traits;
mod objects;
mod people;
mod actions;

use people::{Patient, Nurse, Doctor};
use objects::PatientList;
use actions::{admit_patient, diagnose_patient, \
    prescribe_meds, administer_meds, discharge_patient};
```

2. In our main function, we can now define the two nurses and doctors for our clinic for the day with the following code:

```
fn main() {
    let doctor = Doctor{name: String::from("Torath")};
    let doctor_two = Doctor{name: \
      String::from("Sergio")};
```

```
        let nurse = Nurse{name: String::from("Maxwell")};
        let nurse_two = Nurse{name: \
          String::from("Nathan")};
    }
```

3. We then get our patient list, and it turns out that the four horsemen have turned up for their treatment, as seen in the following code snippet:

```
        let patient_list = PatientList {
            patients: vec![
                Box::new(Patient{name: \
                  String::from("pestilence")}),
                Box::new(Patient{name: \
                  String::from("war")}),
                Box::new(Patient{name: \
                  String::from("famine")}),
                Box::new(Patient{name: \
                  String::from("death")})
            ]
        };
```

4. We then loop through our patients, getting our doctors and nurses to care for them with the following code:

```
        for i in patient_list.patients {
            admit_patient(&i, &nurse);
            diagnose_patient(&i, &doctor);
            prescribe_meds(&i, &doctor_two);
            administer_meds(&i, &nurse_two);
            discharge_patient(&i, &nurse);
        }
```

This is the end of our main function. Running it would give us the following printout:

```
conquest is being admitted
conquest is being diagnosed
conquest is being prescribed medication
conquest is having meds administered
conquest is being discharged
```

```
war is being admitted
. . .
famine is being admitted
. . .
death is being admitted
. . .
```

With this, we have finished our exercise in using traits in Rust. Hopefully, with this, you see the flexibility and decoupling we get when we use traits. However, we must remember that this approach cannot be supported if we were to build an interface with our Python system. If we were to build an interface, this could be done with the following pseudocode:

```
#[pyclass]
pub struct NurseClass {
    #[pyo3(get, set)]
    pub name: String,
    #[pyo3(get, set)]
    pub admin: bool,
    #[pyo3(get, set)]
    pub prescribe: bool,
    #[pyo3(get, set)]
    pub diagnose: bool,
}
#[pymethods]
impl NurseClass {
    #[new]
    fn new(name: String, admin: bool, prescribe: bool,
            diagnose: bool)        Self {
        return Nurse{name, admin, prescribe}
    }
    fn introduce(&self)        Vec<Vec<u64>> {
        println!("hello I'm a Nurse and my name is {}",
            self.name);
    }
}
```

Here, we can see that we swapped the functions in the `ClinicalSkills` trait for attributes. We would be able to pass our `NurseClass` struct with traits into a function which calls the `ClinicalSkills` functions. The results from the `ClinicalSkills` functions can then be passed into the constructor for our `NurseClass` struct. Our `NurseClass` struct can then be passed out to our Python system.

OOP has its merits and should be used when coding in Python. However, Rust has given us a new approach that is flexible and decoupled. It may take a while to get your head around traits, however, they are worth it. It's advised that you keep working with traits in your Rust code to get the advantages of using traits.

Keeping data-parallelism simple with Rayon

In *Chapter 3*, *Understanding Concurrency* we processed our Fibonacci numbers in parallel. While it was interesting to look into concurrency, when we are building our own applications, we should lean on other crates to reduce the complexity of our application. This is where the `rayon` crate comes in. This will enable us to loop through numbers to be calculated and process them in parallel. In order to do this, we initially have to define the crate in the `Cargo.toml` file as seen here:

```
[dependencies]
rayon = "1.5.1"
```

With this, we import this crate in our main.rs file with the following code:

```
extern crate rayon;
use rayon::prelude::*;
```

Then, if we do not import the macros with `use rayon::prelude::*;` our compiler will refuse to compile when we try and turn a standard vector into a parallel iterator. With these macros, we can execute parallel Fibonacci calculations with the following code:

```
pub fn fibonacci_reccursive(n: i32) -> u64 {
    match n {
        1 | 2 => 1,
        _    => fibonacci_reccursive(n - 1) +
                    fibonacci_reccursive(n - 2)
    }
}
fn main() {
    let numbers: Vec<u64> = vec![6, 7, 8, 9, 10].into_par_iter(
```

```
    ).map(
        |x| fibonacci_reccursive(x)
    ).collect();
    println!("{:?}", numbers);
}
```

With this code, we can see that we define a standard Fibonacci number function. We then get a vector of input numbers and convert it into a parallel iterator with the `into_par_iter` function. We then map our Fibonacci function to this parallel iterator. After this, we collect the results. Therefore, printing `numbers` will give us [8, 13, 21, 34, 55]. And that's it! We have coded parallel code, and we have kept it simple with the `rayon` crate. However, we must remember that there is a cost to set up this parallelization. If we were only going to use the numbers in the example, a normal loop would be faster. However, if the numbers and size of the array increase, the benefits of `rayon` start to show. For instance, if we were to have a vector of numbers to be calculated ranging from 6 to 33, we will get the time difference as seen in the following figure:

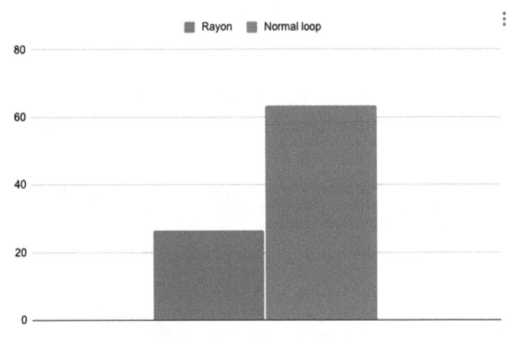

Figure 11.5 – Time taken for a loop 6 -> 33 Fib numbers to be taken in microseconds
[left = Rayon right = normal loop]

With this, we have a simple approach to parallelizing our calculations which will keep our complexity and mistakes down.

Summary

In this chapter, we went over best practices for implementing Rust in our Python systems. We initially started by keeping it simple. We saw that we could leverage the speed of Rust without any setup tools or installing a package thanks to piping data to and from our Rust binary with Python. This is a useful technique to have and is not just limited to Python and Rust. In fact, you can pipe data between any language.

If you are writing a basic program, then data piping should be the first thing you should do. This way, you reduce the number of moving parts and speed up development. A simple Bash script could compile the Rust file and run the process. However, as the program complexity increases, you can go for setup tools and import your Rust code directly into your Python code, utilizing what you covered in this book.

We then moved on to the importance of leveraging Python's object support with a metaclass to lean on Python for our interfaces without Rust packages. Python is a mature language that is very expressive. It makes sense to use the best of Python and the best of Rust when building our packages by using Python for the interfaces and Rust for the calculations. We finally covered how to utilize traits as opposed to forcing Rust to have an object-orientated approach with inheritance via composition. The result is more decoupling and flexibility. Finally, we kept our parallel processing code simple with third-party crates which will increase our productivity and reduce the complexity of our code, and in turn, reduce mistakes.

We have now come to the end of the book. There is always more to learn; however, you now have a full tool belt. You not only have a handle on the fastest memory-safe language that is cutting-edge, but you can also fuse it with the widely used Python language in an efficient way, installing it with `pip`. Not only can you do this for Python scripts, but you can also wrap up Rust extensions in Docker, enabling you to use Rust in Python web applications. Therefore, you do not have to wait for your company and projects to rewrite and adopt Rust. Instead, you can plug Rust into an already established project tomorrow. I am nothing short of excited about what you will do with this in the future.

Further reading

- *Mastering Object-Oriented Python* by *Steven Lott* (2019) (*Packt Publishing*)
- *Mastering Rust* by *Rahul Sharma, Vesa Kaihlavirta* (2018) (*Packt Publishing*)

Index

A

adapter interface
  building, with singleton design
      pattern 160-164
  testing, in Python console 165, 166
adapters
  creating, for package 159, 160
Amdahl's law 95, 96
application database migration system
  setting up 280-283
application programming
      interface (API) 74, 190
asynchronous programming
  with threads 74-82

B

basic Flask application
  building 262, 263
  Docker image, building for 266-268
  entry point, building for 263, 264
  Fibonacci number calculator
      module, building 264, 265
  NGINX service, building 268-270

  Nginx service, connecting 270-273
  Nginx service, running 270-273
Bh script
  building 310

C

calculate functions
  adding, to Python bindings 227
calculate_parameters functions
  building 226, 227
calculate_times functions
  building 226, 227
calculation model
  building, with Rust 301-304
calculation view
  creating, with Rust 304, 305
  updating 291, 292
Cargo
  code, managing with 42-50
catastrophe model
  flow 236
  linear interpolation of distribution 238
catastrophe modeling problem
  breaking down 234-236

Celery
  about 97, 98
  deploying, with Rust 307, 308
  Fibonacci calculation task,
    building for 290
  Rust, fusing into 300, 305, 306
Celery broker
  building, for Flask 288, 289
Celery service
  defining, in Docker 292-295
central processing unit (CPU) 73
clogging 96
code
  managing, with Cargo 42-50
  managing, with crates 42-50
  structuring, over multiple files
    and modules 50-55
command-line tools
  creating, for package 157, 158
comma-separated values (CSV) 328
complex Python objects
  data, extracting from config file 183, 184
  data, processing from Python
    dictionary 179-183
  passing, into Rust 176
  Rust dictionary, returning to
    Python system 185-187
  setup.py file, updating to support
    .yml loading 177, 178
  .yml loading command,
    defining 178, 179
concurrency 72
config loading system
  building 275, 277
continuous integration, setting up
  about 125
  automatic versioning, creating
    for pip package 135-138

dependencies, managing 127-129
deploying onto PyPI, with
    GitHub actions 138-140
GitHub actions, setting up 131-134
GitHub repository, deploying
    manually onto PyPI 126, 127
steps 125
type checking, setting up for
    Python 129, 130
type-checking, with GitHub
    actions 131-135
crates
  code, managing with 42-50
custom Python objects
  attributes, setting 193, 194
  class constructor, defining 197
  class static methods, defining to
    process input numbers 196
  constructing, in Rust 195
  data, adding to PyDict struct 191-193
  inspecting with 187
  module, testing 198-201
  module, wrapping up 198-201
  object, creating for Rust interface 188
  Python class, defining with
    attributes 195, 196
  Python GIL, acquiring in Rust 189, 190
  working with 187

D

database access layer
  applying, to fib calculation view 285-287
  building 277-280
  defining 273, 274
database models
  building 283-285

data parallelism
   with Rayon crate 349-351
deadlock 96, 97
dictionaries
   replacing, with hashmaps 14-18
Docker
   Celery service, defining in 292-295
docker-compose
   PostgreSQL database, defining in 275

E

end-to-end Python package
   building 239, 240
   footprint merging process,
      building 240-243
   package installation instructions,
      building 250-252
   Python construct model, building
      with Pandas 253, 254
   Python interface, building 249
   Python interface, building
      in Rust 247-249
   random event ID generator
      function, building 254
   testing 252, 255
   utilizing 252
   vulnerability merge process,
      building 243-246
Enum 14
environment
   interacting with 65-67
error handling
   in Rust 18-21

F

fib calculation view
   database access layer,
      applying to 285-287
Fibonacci calculation code
   building 114-116
Fibonacci calculation task
   building, for Celery 290
Fibonacci number calculator module
   building 264, 265
Fibonacci Rust code
   building 153-156
Fibonacci sequence 84
Flask
   Celery broker, building for 288, 289
   deploying, with Rust 307, 308
   Rust, fusing into 300
floats 11
footprint merging process
   building 240-243
functions
   used, for passing traits 344, 345

G

garbage collection 6
get_input_vector functions
   building 224, 225
get_parameters functions
   building 224, 225
get_times functions
   building 224, 225
get_weight_matrix functions
   building 223, 224

GitHub repository
   creating 106-109
global interpreter lock (GIL) 73, 187

H

hashmaps
   used, for replacing dictionaries 14-18

I

input/output (I/O) 73
integers 11
interfaces as objects 334-337
inverse_weight_matrix functions
   building 223, 224

J

JavaScript Object Notation (JSON) 37

L

locks 98

M

macros
   about 35
   used, for metaprogramming 34-37
message bus
   building 287, 288
metaclasses 160
mixins 13
mock crates 170

model, in NumPy
   about 213-215
   Python object, building for
      execution 216-219
module interfaces
   building 55-63
   coding documentation, benefits 64
Moore's Law 5
multiple processes
   running 82-95
mutable hashmap 15
mutable variable 11

N

NumPy
   exploring 206
   model, building 213
   using, in Rust 219-221
   vectors, adding in 206-208
   vectors, adding in pure Python 208, 209
   vectors, adding in Rust 209-213
NumPy dependency
   adding, in setup.py file 227
NumPy model
   recreating, in Rust 222, 223

O

Object orientated programming
   (OOP) 338

P

package installation instructions
   building 250-252
parallelism 72

patching 122
pip
  Rust, packaging 146
Polars 257
pool 83
PostgreSQL database
  defining, in docker-compose 275
private GitHub repository
  deploying with 308, 309
processes
  about 74
  customizing 95
pure Python
  vectors, adding 208, 209
pyO3 crate
  used, for building Rust interface 153
Python
  versus Rust 5
  fusing, with Rust 5-8
Python and Rust implementations
  timing, with series of different
      data sizes 255-257
Python bindings
  calculate functions, adding to 227
Python code, packaging in pip module
  command-line interface,
      creating 116-118
  Fibonacci calculation code,
      building 114-116
  steps 113
  unit tests, building 118-124
Python console
  adapter interface, testing 165, 166
Python construct model
  building, with Pandas 253, 254
Python interface
  building 228-249
  building, in Rust 247, 248

Python modules
  using, in Rust 219-221
Python pip module
  setup tools, configuring for 106
Python script
  building to formulate numbers
      for calculation 329
  building, to print calculated
      numbers 331-333
  building, with calculated
      numbers 331-333
Python strings
  passing, in Rust 8, 9

R

race condition 98
rand crate 65
random event ID generator function
  building 254
Rayon crate
  using, in data parallelism 349-351
README file
  defining 111
recursion 84
Rust
  arrays data, managing 12-14
  calculation model, building
      with 301-304
  calculation view, creating 304, 305
  Celery, deploying with 307, 308
  complex Python objects,
      passing into 176
  custom Python objects,
      constructing 195
  dependency, defining on Fibonacci
      number calculation package 301

deploying, in Flask 322, 323
error handling 18-21
Flask, deploying with 307, 308
floats, sizing up 10, 11
fusing, into Celery 300-306
fusing, into Flask 300
fusing, with data access 312, 313
integers, sizing up 10-12
NumPy model, recreating 222, 223
NumPy, using 219-221
packaging, with pip 146
Python modules, using 219-221
Python strings, passing 8, 9
vectors data, managing 12-14
versus Python 5
Rust Fib package installment
    configuring, in Dockerfile 311, 312
Rust file
    building to calculate Fibonacci
        numbers 330
    building, to calculate Fibonacci
        numbers 331
    building, to return calculated
        numbers 330, 331
    building, with numbers 330, 331
Rust, fusing with data access
    database cloning package,
        setting up 313-315
    database connection, defining 319
    database models, autogenerating 317
    database models, configuring 317, 318
    database schema,
        autogenerating 317, 318
    diesel environment, setting up 315
    function, creating for obtaining
        Fibonacci records 320-322

Rust implementation, by data piping
    about 328
    Python script, building to formulate
        numbers for calculation 329
    Python script, building to print
        calculated numbers 332, 333
    Python script, building with
        calculated numbers 332, 333
    Rust file, building to calculate
        Fibonacci numbers 330, 331
    Rust file, building to return
        calculated numbers 330, 331
    Rust file, building with
        numbers 330, 331
Rust interface
    building, with pyO3 crate 153
Rust package
    unit tests, building for 167-170
Rust pip module, steps
    about 146
    Cargo, defining for package 147-149
    gitignore, defining for package 147-149
    Python setup process, configuring
        for package 149-151
    Rust library, installing for
        package 151, 152

S

scope
    defining 26-30
setup.py file
    NumPy dependency, adding 227
setup tools, configuring for
        Python pip module
    about 106
    basic module, defining 111, 112
    basic parameters, defining 109-111

GitHub repository, creating 106-109
README file, defining 111
steps 106
signed integers 11
singleton design pattern
using, to build adapter interface 160-164
speed
comparing, with Python, Rust,
and Numba 170-173
stocks module 50
strings 8
string slice 8
struct
about 13
building 30-34
defining, with traits 341-343
storing, with traits 345

T

threads
about 72, 73
customizing 95
using, in asynchronous
programming 74-82
traits
about 13
defining 339, 340
passing, through functions 344, 345
running 346, 347
used, for defining struct 341-343
used, for storing structs 345
using, as opposed to objects 338, 339
two-dimensional (2D) trajectory 334
typosquatting 113

U

unit tests
about 118
building 118-124
building, for Rust package 167-170
unsigned integers 11

V

variable ownership
about 21, 22
copy 22
immutable borrow 24, 25
move 23, 24
mutable borrow 26
vectors
adding, in NumPy 206-208
adding, in pure Python 208, 209
adding, in pure Rust with
NumPy 209-213
vulnerability merge process
building 243-246

Packt.com

Subscribe to our online digital library for full access to over 7,000 books and videos, as well as industry leading tools to help you plan your personal development and advance your career. For more information, please visit our website.

Why subscribe?

- Spend less time learning and more time coding with practical eBooks and Videos from over 4,000 industry professionals

- Improve your learning with Skill Plans built especially for you

- Get a free eBook or video every month

- Fully searchable for easy access to vital information

- Copy and paste, print, and bookmark content

Did you know that Packt offers eBook versions of every book published, with PDF and ePub files available? You can upgrade to the eBook version at packt.com and as a print book customer, you are entitled to a discount on the eBook copy. Get in touch with us at customercare@packtpub.com for more details.

At www.packt.com, you can also read a collection of free technical articles, sign up for a range of free newsletters, and receive exclusive discounts and offers on Packt books and eBooks.

Other Books You May Enjoy

If you enjoyed this book, you may be interested in these other books by Packt:

Python for Geeks

Muhammad Asif

ISBN: 9781801070119

- Understand how to design and manage complex Python projects
- Strategize test-driven development (TDD) in Python
- Explore multithreading and multiprogramming in Python
- Use Python for data processing with Apache Spark and Google Cloud Platform (GCP)
- Deploy serverless programs on public clouds such as GCP

- Use Python to build web applications and application programming interfaces
- Apply Python for network automation and serverless functions
- Get to grips with Python for data analysis and machine learning

Rust Web Programming

Maxwell Flitton

ISBN: 9781800560819

- Structure scalable web apps in Rust in Rocket, Actix Web, and Warp
- Apply data persistence for your web apps using PostgreSQL
- Build login, JWT, and config modules for your web apps
- Serve HTML, CSS, and JavaScript from the Actix Web server
- Build unit tests and functional API tests in Postman and Newman
- Deploy the Rust app with NGINX and Docker onto an AWS EC2 instance

Packt is searching for authors like you

If you're interested in becoming an author for Packt, please visit `authors.packtpub.com` and apply today. We have worked with thousands of developers and tech professionals, just like you, to help them share their insight with the global tech community. You can make a general application, apply for a specific hot topic that we are recruiting an author for, or submit your own idea.

Share Your Thoughts

Now you've finished *Speed Up your Python with Rust*, we'd love to hear your thoughts! Scan the QR code below to go straight to the Amazon review page for this book and share your feedback or leave a review on the site that you purchased it from.

`https://packt.link/r/1-801-81144-X`

Your review is important to us and the tech community and will help us make sure we're delivering excellent quality content.

www.ingramcontent.com/pod-product-compliance
Lightning Source LLC
Chambersburg PA
CBHW062046050326
40690CB00016B/3002